ORIGO
STEPPING STONES
CORE MATHEMATICS

SENIOR AUTHORS
James Burnett
Calvin Irons

PROGRAM CONSULTANTS
Diana Lambdin
Frank Lester, Jr.
Kit Norris

CONTRIBUTING AUTHORS
Debi DePaul
Peter Stowasser
Allan Turton

PROGRAM EDITORS
James Burnett
Beth Lewis
Donna Richards

ORIGO EDUCATION

PRACTICE BOOK

INTRODUCTION

ORIGO STEPPING STONES

The *ORIGO Stepping Stones* program has been created to provide a smarter way to teach and learn mathematics. It has been developed by a team of experts to provide a world-class math program.

PRACTICE BOOK

Regular and meaningful practice is a hallmark of *ORIGO Stepping Stones*. Each module in this book has pages that practice content previously learned to maintain concepts and skills, and pages that practice computation to promote fluency.

PERFORATED PAGES
The pages of this book have been perforated for your convenience.

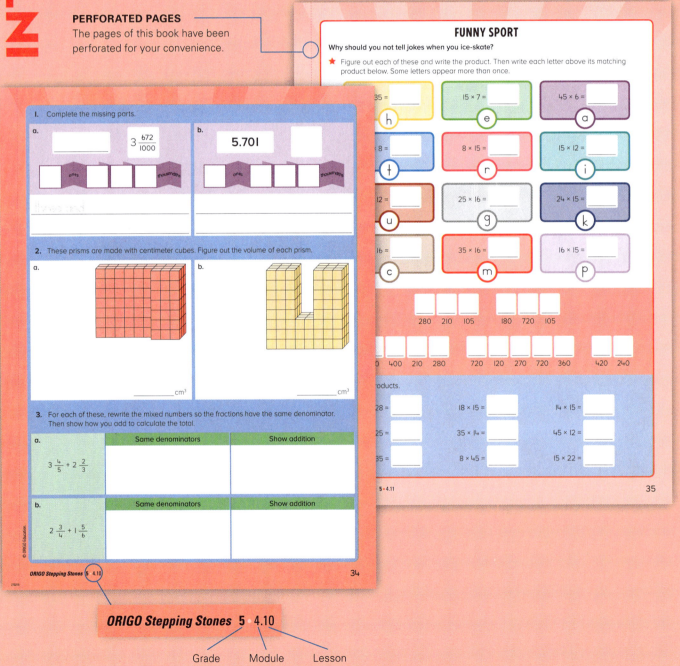

ORIGO Stepping Stones 5 • 4.10
Grade — Module — Lesson

STUDENT JOURNAL

The student journal provides a double-page spread for each lesson in the *ORIGO Stepping Stones* program for Grade 5. Each spread includes guided discussion of enquiry, questions based on the discussion, and a final question that puts a little twist on the content to promote higher-order thinking skills.

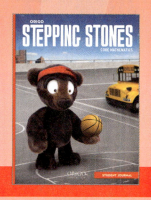

NOTES FOR HOME

Each book is one component of a comprehensive teaching program. Together they are a collection of consolidation and practice pages from lessons in the *ORIGO Stepping Stones* program.

Class teachers will decide which pages suit individual needs. So students might not complete every page in these books. For more information about the program, visit www.origoeducation.com/steppingstones.

ADDITIONAL RESOURCES – PRINT

The Number Case provides teachers with ready-made resources that are designed to develop students' understanding of number.

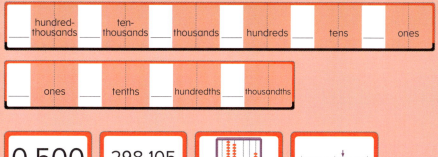

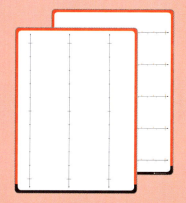

ADDITIONAL RESOURCES (ONLINE CHANNELS)

These are some of the innovative teaching channels integrated into the teacher's online program.

ORIGO MathEd
Professional learning sessions

Flare
Interactive whiteboard tools

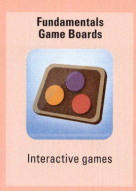

Fundamentals Game Boards
Interactive games

INTRODUCTION

1. Each square is one whole. Read the fraction name and shade the squares to match. Write the matching decimal on the open expander.

   a. one and four-tenths

   b. two and seven-tenths

2. The distance between each whole number is one whole. Write the decimal fraction that is shown by each arrow. Think carefully before you write.

   a. _____ b. _____ c. _____ d. _____

   3 ————————————————————————— 4

   e. _____ f. _____ g. _____ h. _____

3. Complete the missing parts.

   a.
   H T O | H T O
   Thousands | Ones

   three hundred nineteen thousand four hundred twenty-six

   b.
   H T O | H T O
   Thousands | Ones

   706,412

ORIGO Stepping Stones 5 • 1.2

SIMON SAYS

★ To reveal a fun fact, figure out each of these and write the total. Write each letter above its matching total at the bottom of the page. Some letters appear more than once.

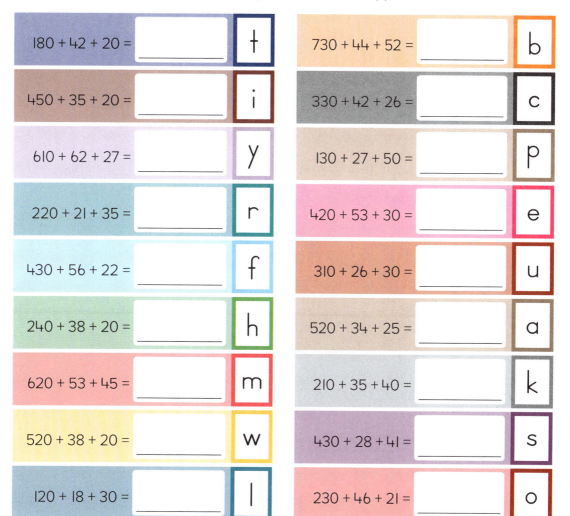

180 + 42 + 20 =	t	730 + 44 + 52 =	b
450 + 35 + 20 =	i	330 + 42 + 26 =	c
610 + 62 + 27 =	y	130 + 27 + 50 =	p
220 + 21 + 35 =	r	420 + 53 + 30 =	e
430 + 56 + 22 =	f	310 + 26 + 30 =	u
240 + 38 + 20 =	h	520 + 34 + 25 =	a
620 + 53 + 45 =	m	210 + 35 + 40 =	k
520 + 38 + 20 =	w	430 + 28 + 41 =	s
120 + 18 + 30 =	l	230 + 46 + 21 =	o

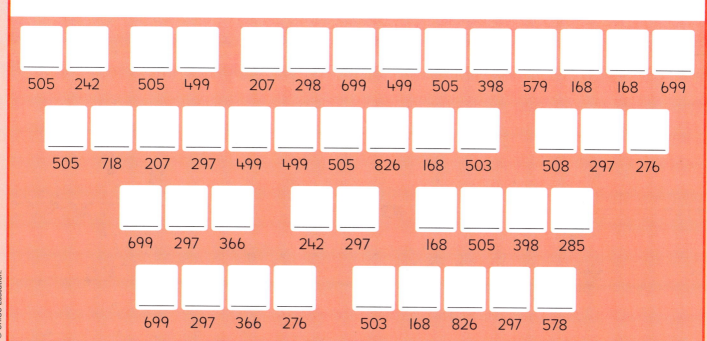

505 242   505 499   207 298 699 499 505 398 579 168 168 699

505 718 207 297 499 499 505 826 168 503   508 297 276

699 297 366   242 297   168 505 398 285

699 297 366 276   503 168 826 297 578

1. Complete these.

a.
```
   3 4 6 1 8
 + 1 6 7 4 3
 _____
```

b.
```
   1 2 7 0 4
 + 7 6 5 8 9
 _____
```

c.
```
   4 7 6 0
   3 1 0 8
 + 2 2 9 5
 _____
```

d.
```
   4 7 9 5 1
 - 1 5 3 6 7
 _____
```

e.
```
   3 2 0 7 5
 - 1 6 9 5 2
 _____
```

f.
```
   2 7 3 0 6
 - 1 5 2 9 8
 _____
```

2. Shade each large square to show the fraction. Then complete the equivalence statement.

a. Shade $\frac{1}{2}$ of each.

 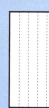

$\frac{1}{2} = \frac{}{10} = 0.\underline{}$

b. Shade $\frac{2}{5}$ of each.

 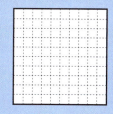

$\frac{2}{5} = \frac{}{100} = 0.\underline{}$

3. Write an expression to explain what each of these means.

a. 10^2 _____

b. 10^3 _____

c. 10^4 _____

d. 10^5 _____

e. 10^6 _____

1. Complete the missing parts.

a.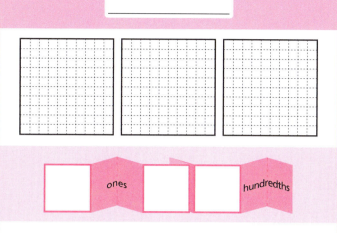

two and sixteen hundredths

b.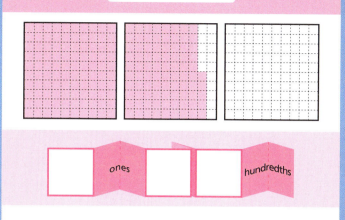

2. Calculate the total distance for each of these.

a. 3.2 km + 4.5 km = ___ km

b. 5.1 km + 2.7 km = ___ km

c. 2.6 km + 1.3 km = ___ km

d. 2.3 km + 6.4 km = ___ km

e. 4.5 km + 2.4 km = ___ km

f. 4.6 km + 2.2 km = ___ km

3. Read the number. Then write the matching numeral.

a. five million twelve thousand one hundred sixty-two

b. seven million five hundred six thousand two hundred fifteen

c. one million four hundred nine thousand

d. nine million one hundred seventy-one thousand four hundred forty-eight

e. six million twenty thousand nine hundred one

CROSS NUMBER

★ Write the differences in the puzzle grid below.

Across

a. 792 − 550
c. 283 − 110
e. 883 − 360
f. 785 − 340
h. 468 − 230
j. 873 − 450
l. 585 − 320
n. 393 − 150
o. 292 − 120
p. 976 − 220

Down

a. 574 − 350
b. 395 − 140
c. 372 − 240
d. 768 − 450
g. 962 − 530
i. 486 − 130
j. 681 − 250
k. 562 − 240
l. 547 − 310
m. 866 − 330

1. Each square is one whole. Draw lines and shade parts to show the first fraction. Then draw extra lines to help you identify the equivalent fraction.

a.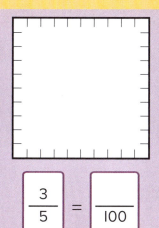
$\frac{3}{5} = \frac{}{100}$

b.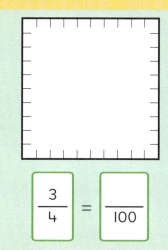
$\frac{3}{4} = \frac{}{100}$

c.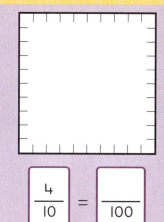
$\frac{4}{10} = \frac{}{100}$

2. Write the total cost. Show your thinking.

a. $2.55 $3.20

$ _____

b. $1.42 $5.35

$ _____

c. $3.15 $2.70

$ _____

d. $4.25 $2.44

$ _____

3. Write the number shown by each arrow.

a. _____   b. _____   c. _____   d. _____

2,120,000   2,130,000   2,140,000

e. _____   f. _____   g. _____   h. _____

1. These rectangles have been split into two parts to make it easier to divide. Write the missing numbers. Then complete the equation.

a.
52 ÷ 4 = _____

4 | 40 | 12

_____ + _____

b.
75 ÷ 5 = _____

5 | 50 | 25

_____ + _____

2. Show each decimal fraction as the sum of three numbers.

a. 1.56 = __1__ + __0.5__ + __0.06__

b. 2.98 = _____ + _____ + _____

c. 1.07 = _____ + _____ + _____

d. 1.45 = _____ + _____ + _____

e. 0.51 = _____ + _____ + _____

f. 2.50 = _____ + _____ + _____

3. Use a pattern to help you write the products.

a.
	Th	H	T	Ones
12 × 2 =				
12 × 20 =				
120 × 20 =				

b.
	Th	H	T	Ones
13 × 4 =				
13 × 40 =				
130 × 40 =				

c.
	Th	H	T	Ones
21 × 4 =				
21 × 40 =				
210 × 40 =				

d.
	Th	H	T	Ones
15 × 3 =				
15 × 30 =				
150 × 30 =				

e.
	Th	H	T	Ones
14 × 6 =				
14 × 60 =				
140 × 60 =				

f.
	Th	H	T	Ones
32 × 3 =				
32 × 30 =				
320 × 30 =				

RACE TRACK

★ Figure out and write the answers to these as fast as you can. Use the classroom clock to time yourself.

Time Taken:

1. Divide each part and complete the equation.

a.
448 ÷ 4 = _____

| 4 | 400 | 40 | 8 |

_____ + _____ + _____

b.
630 ÷ 6 = _____

| 6 | 600 | 30 |

_____ + _____

2. Loop the bag of items that weighs **more than 1 lb** in total.
Remember there are 16 ounces in 1 pound.

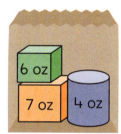

3. Double one number and halve the other to make a problem that is easier to solve.
Then write the product.

a. 35 × 14
_____ × _____
35 × 14 = _____

b. 18 × 15
_____ × _____
18 × 15 = _____

c. 45 × 8
_____ × _____
45 × 8 = _____

d. 15 × 26
_____ × _____
15 × 26 = _____

e. 35 × 24
_____ × _____
35 × 24 = _____

f. 28 × 15
_____ × _____
28 × 15 = _____

g. 25 × 14
_____ × _____
25 × 14 = _____

h. 16 × 45
_____ × _____
16 × 45 = _____

i. 35 × 18
_____ × _____
35 × 18 = _____

1. Figure out each partial product. Then write the total of the four products.

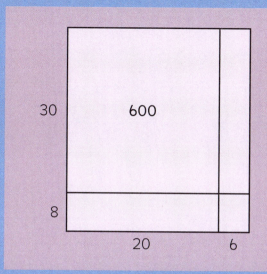

38 × 26

30 × __20__ = _____

30 × __6__ = _____

8 × __20__ = _____

8 × __6__ = _____

Total _____

2. Write the matching number or number name.

a. three hundred six million twenty-five thousand nine hundred two

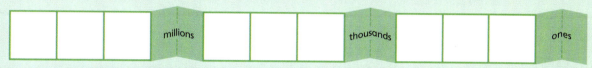

b.

| | 1 | 2 | millions | 7 | 5 | 0 | thousands | 6 | 1 | 9 | ones |

_____

_____

_____

3. Draw an arrow from each fraction and mixed number to show its position on one number line.

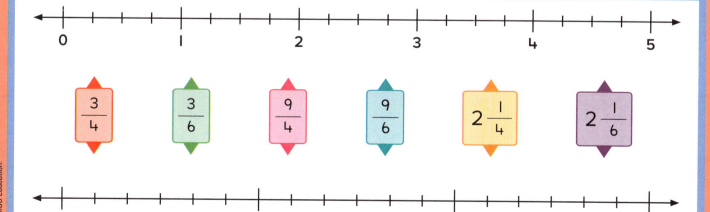

MORE! MORE! MORE!

The more there is of me, the less you see. What am I?

★ Figure out each of these and write the product. Find each product in the grid below and cross out the letter above. Then write the remaining letters at the bottom of the page.

5 × 8 × 4 = ____

2 × 7 × 5 = ____

4 × 5 × 2 = ____

5 × 8 × 7 = ____

7 × 6 × 5 = ____

5 × 6 × 2 = ____

6 × 5 × 0 = ____

9 × 5 × 4 = ____

5 × 9 × 6 = ____

4 × 6 × 5 = ____

2 × 5 × 9 = ____

5 × 5 × 7 = ____

9 × 1 × 5 = ____

8 × 5 × 9 = ____

4 × 7 × 5 = ____

5 × 6 × 5 = ____

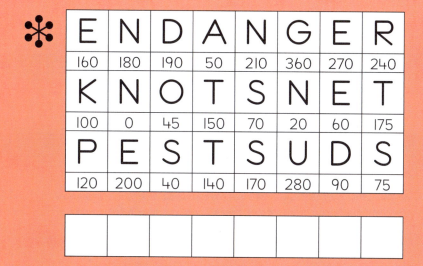

Write the letters in order from the ✳ to the bottom-right corner.

1. Write the dimensions around the rectangle. Figure out the product for each part. Then add the four partial products to find the total.

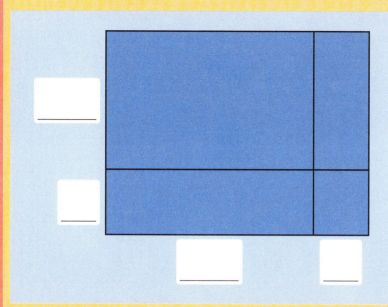

37 × 44

____ × ____ = ____

____ × ____ = ____

____ × ____ = ____

____ × ____ = ____

Total ____

2. Draw an arrow to show where you think each number is on the number line.

$\frac{1}{2}$ million $\frac{3}{4}$ million $1\frac{1}{2}$ million $1\frac{3}{4}$ million

0 — 2,000,000

250,000   1,000,000   1,250,000   1,750,000

3. For each of these, write the fraction and then write four fractions that are equivalent.

a. three-fifths

b. three-fourths

c. eight-sixths

1. Calculate the difference. Draw jumps on the number line to show your thinking.

a. $3\frac{5}{6} - 2\frac{2}{6} = $

b. $4\frac{4}{5} - 2\frac{1}{5} = $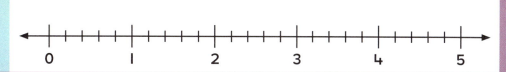

2. Calculate the products. Show the steps you use.

a. $25 \times 18 = $ _____

b. $45 \times 14 = $ _____

c. $28 \times 35 = $ _____

3. For each pair of fractions, write equivalent fractions that have denominators the same. Write the missing factors to show your thinking.

a.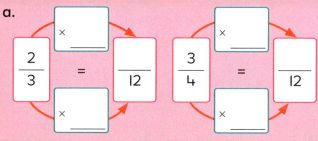

b.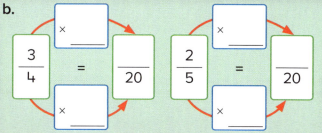

c.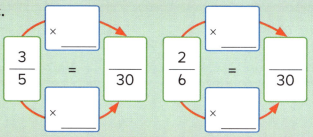

d.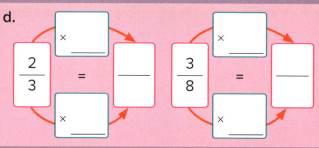

AUSSIE JUMPER

What is another name for the female red kangaroo?

★ Figure out and write the missing factors.
Then write each letter above its matching factor below.

4 × ___ = 36 (l)	2 × ___ = 48 (l)	2 × ___ = 150 (h)
5 × ___ = 35 (e)	2 × ___ = 90 (e)	4 × ___ = 88 (e)
2 × ___ = 70 (i)	5 × ___ = 150 (t)	4 × ___ = 60 (b)
2 × ___ = 24 (f)	4 × ___ = 200 (r)	4 × ___ = 100 (u)

___ ___ ___   ___ ___ ___ ___   ___ ___ ___ ___ ___
30 75 22   15 9 25 7   12 24 35 45 50

Write these missing factors as fast as you can.

5 × ___ = 100   2 × ___ = 180   4 × ___ = 120

2 × ___ = 110   5 × ___ = 55   2 × ___ = 44

4 × ___ = 48   5 × ___ = 45   4 × ___ = 84

1. Each large shape is one whole. Complete each equation.

a. 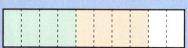 $2 \times \frac{4}{10} =$ _____

b. 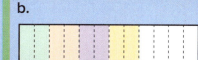 $4 \times \frac{2}{12} =$ _____

c. $5 \times \frac{1}{6} =$ _____

d. 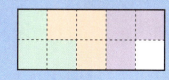 $3 \times \frac{3}{10} =$ _____

e. 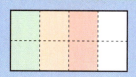 $3 \times \frac{2}{8} =$ _____

2. Break **one** or **both** numbers into two factors to make it easier to multiply. Then write the matching equation.

a. 28 × 25 is the same as _____

b. 45 × 12 is the same as _____

c. 55 × 16 is the same as _____

d. 34 × 15 is the same as _____

3. Compare the fractions in each pair to a benchmark such as $\frac{1}{2}$ or 1. Then write **<** or **>** to make a true statement. Write your thinking.

a. $\frac{2}{7}$ ◯ $\frac{4}{9}$ _____

b. $\frac{5}{6}$ ◯ $\frac{7}{8}$ _____

c. $\frac{3}{4}$ ◯ $\frac{7}{9}$ _____

ORIGO Stepping Stones 5 2.8

1. Complete the missing numbers to calculate each product.

a. $4 \times 2\frac{3}{6}$

$(4 \times \boxed{}) + (4 \times \frac{\boxed{}}{})$

$\boxed{} + \frac{\boxed{}}{} = \boxed{}$

b. $3 \times 2\frac{5}{8}$

$(3 \times \boxed{}) + (3 \times \frac{\boxed{}}{})$

$\boxed{} + \frac{\boxed{}}{} = \boxed{}$

2. Figure out each product. Show the steps you use.

a. $68 \times 3 =$ _____

b. $74 \times 6 =$ _____

c. $47 \times 4 =$ _____

d. $59 \times 5 =$ _____

e. $64 \times 9 =$ _____

f. $83 \times 6 =$ _____

3. Each large square is one whole. Shade parts to show the decimal fraction. Then write the matching common fraction.

a.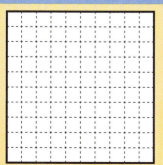

0.35 is the same as $\frac{\boxed{}}{100}$

b.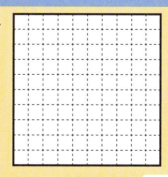

0.05 is the same as $\frac{\boxed{}}{}$

c.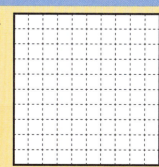

0.5 is the same as $\frac{\boxed{}}{}$

LUNCH TIME

Why can any hamburger run a mile in less than four minutes?

★ Figure out each of these and write the total. Then write each letter above its matching total at the bottom of the page.

$3.25 + $4.40 = $ _____ u
$5.15 + $2.80 = $ _____ a
$1.35 + $2.30 = $ _____ c
$3.40 + $3.15 = $ _____ t
$3.25 + $5.50 = $ _____ a
$4.24 + $1.40 = $ _____ e
$2.35 + $3.15 = $ _____ o
$2.41 + $6.35 = $ _____ s
$1.22 + $3.35 = $ _____ f
$6.74 + $1.14 = $ _____ o

$5.43 + $1.36 = $ _____ i
$1.34 + $7.12 = $ _____ i
$2.55 + $2.14 = $ _____ e
$2.64 + $4.05 = $ _____ d
$6.51 + $3.27 = $ _____ s
$5.32 + $4.33 = $ _____ s
$3.14 + $2.51 = $ _____ b
$3.30 + $5.25 = $ _____ t
$2.71 + $2.28 = $ _____ f

___ ___ ___ ___ ___ ___ ___   ___ ___
$5.65 $5.64 $3.65 $8.75 $7.65 $8.76 $4.69 $6.79 $6.55

___ ___   ___ ___ ___ ___   ___ ___ ___ ___
$8.46 $9.78 $4.99 $7.95 $9.65 $8.55 $4.57 $5.50 $7.88 $6.69

1. Complete the diagrams to show the equivalent fractions.

a. 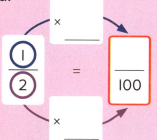   b. 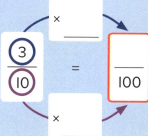   c. 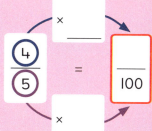   d.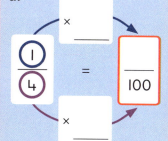

2. Use a method you like to calculate these in your head.

a. 32 × 25 = _____   b. 36 × 25 = _____   c. 48 × 25 = _____

d. 28 × 25 = _____   e. 16 × 25 = _____   f. 24 × 25 = _____

g. 18 × 50 = _____   h. 26 × 50 = _____   i. 44 × 50 = _____

j. 62 × 50 = _____   k. 32 × 50 = _____   l. 34 × 50 = _____

3. Each large square is one whole. Shade parts of the whole to match the fraction on the expander.

a. 0 ones 2 tenths 5 hundredths 5 thousandths

b. 0 ones 7 tenths 1 hundredths 2 thousandths

1. Look at these digit cards.

Use each digit once to make these.

a. the **greatest** and **least** numbers

greatest _____ least _____

b. the greatest **even** number

_____

c. any three numbers that are between 7,500,000 and 8,000,000

_____ _____ _____

2. Rewrite each mixed number as an improper fraction. Show your thinking.

a. $3\frac{4}{5}$ is the same as ▭

b. $6\frac{7}{10}$ is the same as ▭

3. Estimate each product. Then use the standard multiplication algorithm to calculate the exact product.

a. Estimate
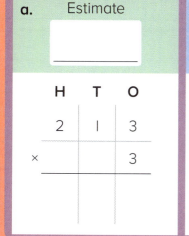

b. Estimate
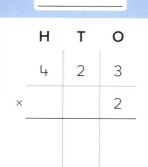

c. Estimate

d. Estimate
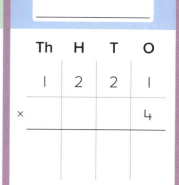

GROWING UP

How often do brain cells regrow?

★ Figure out each of these and write the answer. Find the answer in the grid below and cross out the letter above. Then write the remaining letters at the bottom of the page.

380 − 210 = 170	460 + 280 = 740	630 − 440 = 190
560 + 370 = 930	410 − 150 = 260	550 + 180 = 730
450 − 170 = 280	290 + 560 = 850	420 − 280 = 140
340 + 180 = 520	850 − 370 = 480	280 + 360 = 640
660 − 150 = 510	240 + 460 = 700	750 − 340 = 410
170 + 430 = 600	520 − 180 = 340	260 + 510 = 770
570 + 390 = 960	350 − 230 = 120	450 + 380 = 830
780 − 530 = 250	280 + 470 = 750	430 − 280 = 150
460 + 480 = 940		

O	N	C	E	T	W	E	N	T	Y
930	270	700	170	520	960	710	260	410	480
S	E	V	E	R	A	L	T	W	O
250	730	610	740	510	120	280	600	150	830
O	F	T	E	N	F	O	R	T	Y
340	850	750	360	640	940	190	580	140	770

Write the letters in order from the ✱ to the bottom-right corner.

N	E	V	E	R

1. Calculate the area. Use a pattern to help you.

a.

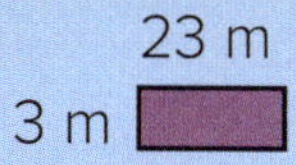
23 m
3 m

3 × 23 = ______ m²

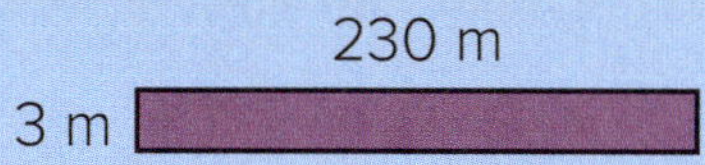
230 m
3 m

3 × 230 = ______ m²

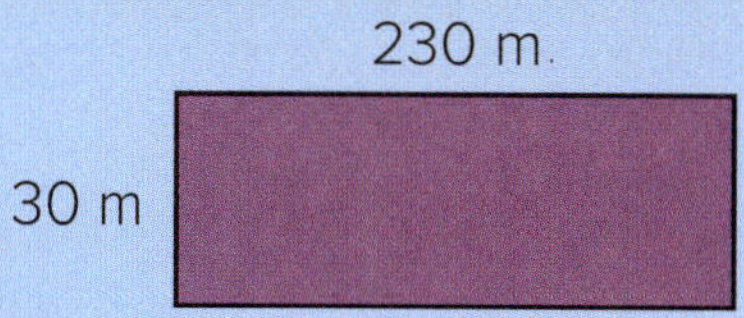
230 m
30 m

30 × 230 = ______ m²

b.

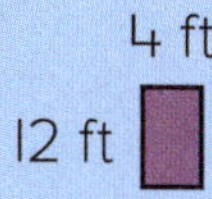
4 ft
12 ft

12 × 4 = ______ ft²

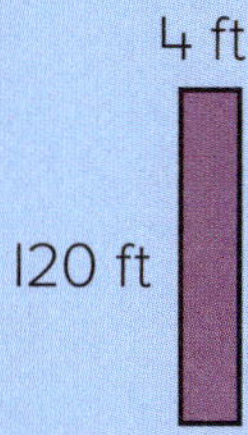
4 ft
120 ft

120 × 4 = ______ ft²

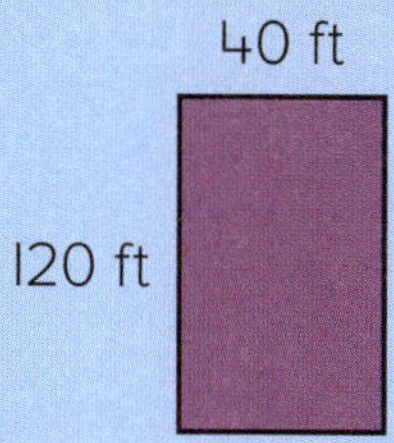
40 ft
120 ft

120 × 40 = ______ ft²

2. For each of these, shade parts to show the decimal fraction.
Then write the matching common fraction.

a.

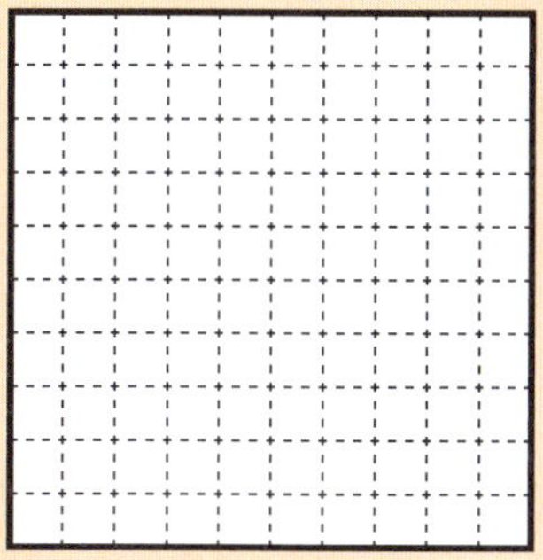

0.09 is the same as ——

b.

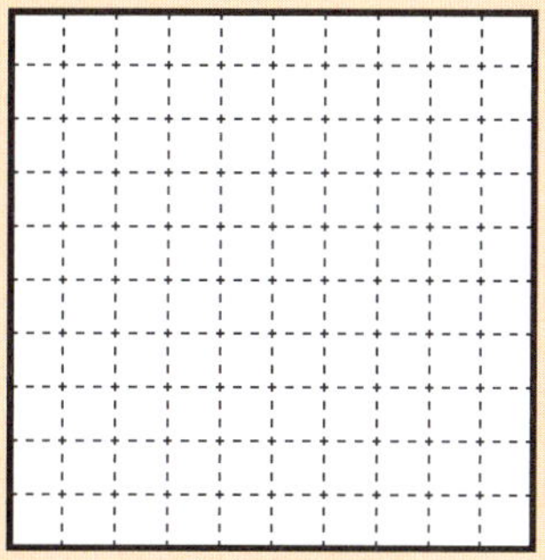

0.19 is the same as ——

c.

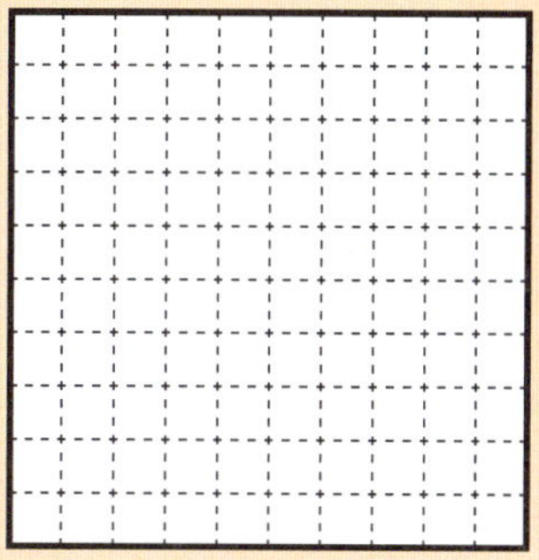

0.9 is the same as ——

3. Write your estimate. Then use the standard multiplication algorithm to calculate the exact product.

a. Estimate ______

	H	T	O
	3	1	8
×			3

b. Estimate ______

	Th	H	T	O
	4	0	8	5
×				6

c. Estimate ______

	Th	H	T	O
	3	1	7	8
×				4

© ORIGO Education.

1. Calculate the perimeter of each frame.

a.

16 in

8 in

2 × 16 = _____

2 × 8 = _____

Perimeter _____ in

b.

6 in

14 in

2 × 6 = _____

2 × 14 = _____

Perimeter _____ in

2. Complete the missing parts.

a.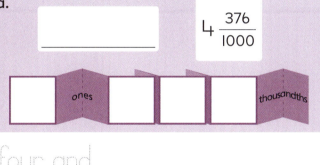

$4\frac{376}{1000}$

four and _____

b.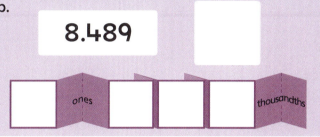

8.489

_____

3. Use the standard multiplication algorithm to calculate the exact product. Then estimate the product to check that your answer makes sense.

a.
```
    2 1 5
  ×   3 4
  -------
```

b.
```
    5 8 6
  ×   3 2
  -------
```

c.
```
    3 6 2
  ×   4 5
  -------
```

DID YOU KNOW?

★ Figure out each of these and write the answer. Then write each letter above its matching answer at the bottom of the page. Some letters appear more than once.

185 ÷ 5 = **37** — l	224 ÷ 4 = **56** — s	522 ÷ 3 = **174** — p
501 ÷ 3 = **167** — n	365 ÷ 5 = **73** — t	316 ÷ 4 = **79** — u
252 ÷ 4 = **63** — o	423 ÷ 3 = **141** — g	225 ÷ 5 = **45** — r
485 ÷ 5 = **97** — c	548 ÷ 4 = **137** — y	414 ÷ 3 = **138** — v
312 ÷ 3 = **104** — a	375 ÷ 5 = **75** — i	568 ÷ 4 = **142** — w
213 ÷ 3 = **71** — h	328 ÷ 4 = **82** — b	132 ÷ 3 = **44** — e

E L E C T R I C   E E L S
44 37 44 97 73 45 75 97   44 44 37 56

S T O R E   E N O U G H
56 73 63 45 44   44 167 63 79 141 71

E L E C T R I C I T Y I N
44 37 44 97 73 45 75 97 75 73 137   75 167

T H E I R   T A I L S   T O
73 71 44 75 45   73 104 75 37 56   73 63

L I G H T   U P   T W E L V E
37 75 141 71 73   79 174   73 142 44 37 138 44

L I G H T   B U L B S
37 75 141 71 73   82 79 37 82 56

1. Draw the reflection on the other side of the dashed line.

a.
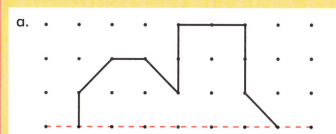

b.
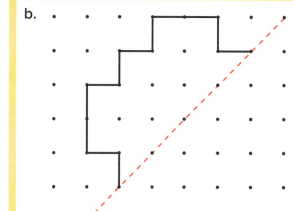

2. Complete the missing parts.

a.
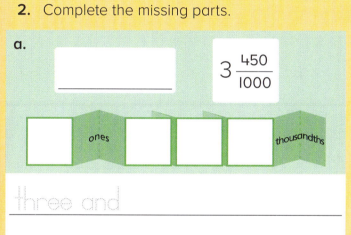

$3\frac{450}{1000}$

ones ... thousandths

three and _____

b.

1.615

ones ... thousandths

3. Use the standard multiplication algorithm to calculate the exact product. Then estimate the product to check that your answer makes sense.

a.
```
    2 0 7 4
  ×     4 2
```

b.
```
      3 2 8
  ×   4 0 5
```

c.
```
      3 2 9
  ×   2 4 7
```

1. The distance between each whole number is one whole. Draw a line to join each numeral to its approximate position on the number line. Be as accurate as possible.

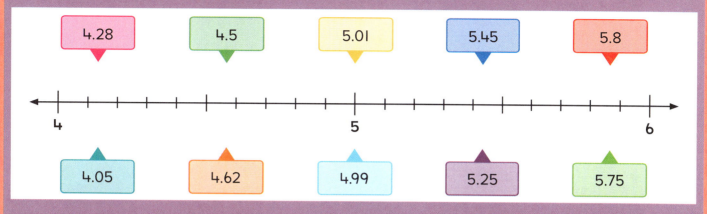

2. Write the decimal fraction that is shown by each arrow.

a.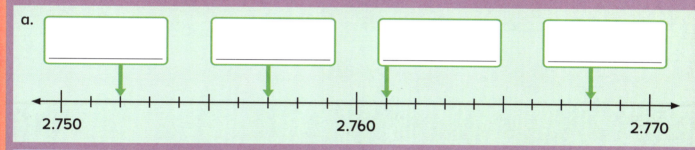

b.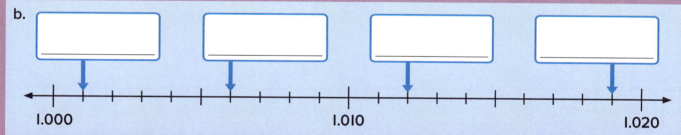

3. Solve these word problems. Show your thinking.

a. A school bought 24 new computers for $1,238 each. What was the total cost?

$ _____

b. A roller coaster carries 32 people. It makes 186 rides each day. What is the greatest number of people who could ride the roller coaster in one day?

_____

DOGS AND DOGS

What is the world's largest dog?

★ Figure out each of these and write the product. Then write each letter above its matching product at the bottom of the page.

265 × 3 = _____ e
3 × 155 = _____ i
4 × 125 = _____ o
185 × 5 = _____ o
3 × 225 = _____ h
3 × 175 = _____ s
245 × 4 = _____ n
115 × 5 = _____ t
135 × 3 = _____ i
165 × 4 = _____ l
5 × 195 = _____ h
3 × 245 = _____ d
175 × 4 = _____ h
155 × 5 = _____ w
315 × 3 = _____ r
4 × 215 = _____ u
5 × 145 = _____ f

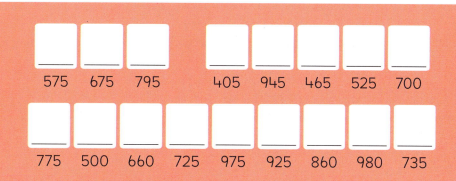

575 675 795   405 945 465 525 700

775 500 660 725 975 925 860 980 735

1. Use what you know about equivalence to calculate each total.

a. $\frac{3}{10} + \frac{45}{100} =$ ☐

b. $\frac{6}{10} + \frac{9}{100} =$ ☐

c. $\frac{1}{10} + \frac{57}{100} =$ ☐

d. $\frac{7}{10} + \frac{15}{100} =$ ☐

e. $3\frac{2}{10} + \frac{30}{100} =$ ☐

f. $1\frac{1}{10} + 2\frac{10}{100} =$ ☐

2. Write an expression that uses decimal fractions to show the parts of these numbers.

a.  3 + 0.6 + _____

b.  _____

c.  _____

d. 5.007 _____

3. Place base-10 ones blocks on the base picture. Build up the number of layers to match the data in the table. Then complete the table.

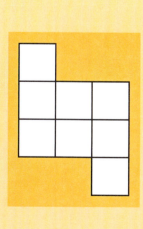

Number of cubes in base	Number of layers	Total number of cubes
8	1	8
8	2	
8	3	
8	4	
8	6	
8	10	

1. Draw beads on each abacus to represent the number.

a. 647,329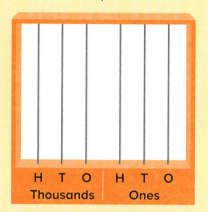

b. 405,612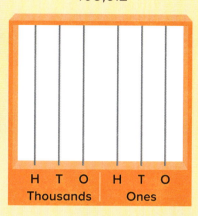

c. 329,060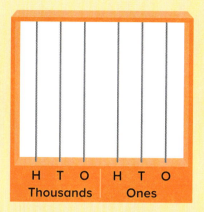

2. Estimate each product. Then use the standard multiplication algorithm to calculate the exact answer.

a. Estimate _____

H	T	O
4	1	6
×		7

b. Estimate _____

Th	H	T	O
4	5	7	2
×			3

c. Estimate _____

Th	H	T	O
8	0	5	9
×			6

3. Write fractions to complete true equations. The fractions in each equation should have the **same denominator**. You can use the number line to help.

a. $\frac{4}{3}$ = ☐ + ☐

b. $2\frac{1}{3}$ = ☐ + ☐

c. $1\frac{2}{3}$ = ☐ + ☐

d. ☐ + ☐ = $\frac{4}{5}$

e. ☐ + ☐ = $2\frac{3}{5}$

f. ☐ + ☐ = $1\frac{4}{5}$

ORIGO Stepping Stones 5 · 4.2

BELIEVE IT OR NOT

★ To reveal a fantastic fact, figure out each of these and write the product. Write each letter above its matching product at the bottom of the page.

5 × 11 = 55	**h**	5 × 83 = 415	**o**
5 × 49 = 245	**i**	5 × 23 = 115	**t**
5 × 35 = 175	**f**	5 × 65 = 325	**l**
5 × 37 = 185	**n**	5 × 17 = 85	**c**
5 × 33 = 165	**d**	5 × 41 = 205	**y**
5 × 39 = 195	**s**	5 × 19 = 95	**w**
5 × 47 = 235	**e**	5 × 27 = 135	**b**

Some letters appear more than once.

T	H	E		B	L	O	O	D
115	55	235		135	325	415	415	165

O	F		I	N	S	E	C	T	S
415	175		245	185	195	235	85	115	195

I	S		Y	E	L	L	O	W
245	195		205	235	325	325	415	95

1. Write the matching number or number name.

a. 412,306,270 _____

b. three hundred five million sixty-two thousand nine hundred three _____

2. Use the standard multiplication algorithm to calculate the exact product. Then estimate the product to check your answer makes sense.

a.
```
      4 8 1
  ×     4 6
  ─────────
```

b.
```
      2 5 9
  ×     3 8
  ─────────
```

c.
```
    5 3 0 5
  ×   2 6 7
  ─────────
```

3. For each of these, rewrite the equation so that the denominators are the same. Use the number line to help. Then write the total.

a. $\dfrac{1}{4} + \dfrac{5}{8} =$ ▭

b. $\dfrac{7}{16} + \dfrac{2}{8} =$ ▭

c. $\dfrac{3}{16} + \dfrac{2}{4} =$ ▭

Number line: 0, 1/16, 1/8, 1/4, ... , 1

1. Break **one** or **both** numbers into two factors to make it easier to multiply. Then write the matching equation.

a. 16 × 35 is the same as _____

b. 14 × 55 is the same as _____

c. 24 × 15 is the same as _____

d. 36 × 45 is the same as _____

2. Complete the table to show the total number of centimeter cubes in each prism.

Dimensions of the base (cm)	Number of layers	Total number of centimeter cubes	Dimensions of the base (cm)	Number of layers	Total number of centimeter cubes
6 × 4	1		7 × 5	2	
6 × 4	2		7 × 5	4	
6 × 4	5		7 × 5	6	
6 × 4	7		7 × 5	8	
6 × 4	10		7 × 5	10	

3. Show how you would figure out each total in two ways.

a. $2\frac{2}{3} + 5\frac{1}{6}$

Use improper fractions	Use mixed numbers

b. $3\frac{1}{3} + 4\frac{7}{12}$

Use improper fractions	Use mixed numbers

ORIGO Stepping Stones 5 · 4.6

THAT'S A FACT

★ To reveal an amazing fact, figure out each of these and write the total. Then write the each letter above its matching total at the bottom of the page.

70 + 344 =	l	538 + 80 =	n
493 + 90 =	i	60 + 685 =	e
50 + 286 =	a	393 + 40 =	s
582 + 70 =	a	471 + 50 =	s
90 + 684 =	a	80 + 249 =	r
337 + 90 =	i	579 + 60 =	l
466 + 60 =	o	687 + 80 =	s
70 + 299 =	c	50 + 365 =	a
555 + 90 =	b	474 + 60 =	r
40 + 671 =	b	30 + 282 =	k
377 + 70 =	p	523 + 90 =	k

652   447   526   414   774   534   711   745   415   329   433

521   613   427   618   583   767   645   639   336   369   312

1. Complete these.

a.
```
    2 6 3 4 7
+   1 3 5 9 5
_____
```

b.
```
    3 5 0 6
    7 2 8 1
+   1 4 3 0
_____
```

c.
```
    3 1 4 6 2
-   1 5 1 7 5
_____
```

d.
```
    4 0 7 5 9
-   2 3 0 6 4
_____
```

2. Calculate the volume of each prism. Then write an equation to show the order that you multiplied the dimensions.

a.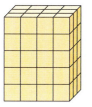

_____ cm³

b.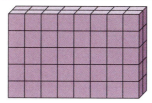

_____ cm³

c.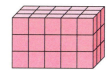

_____ cm³

3. Figure out the totals using improper fractions and then mixed numbers.

a. $3\frac{2}{5} + 1\frac{1}{3}$

Use improper fractions	Use mixed numbers

b. $4\frac{2}{6} + 2\frac{2}{4}$

Use improper fractions	Use mixed numbers

1. Complete the missing parts.

a.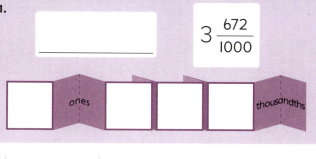

three and _____

b.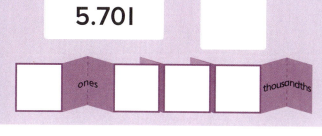

2. These prisms are made with centimeter cubes. Figure out the volume of each prism.

a.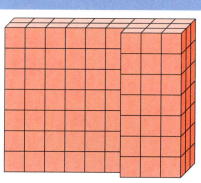

_____ cm³

b.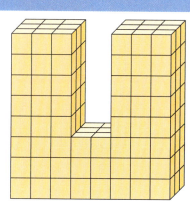

_____ cm³

3. For each of these, rewrite the mixed numbers so the fractions have the same denominator. Then show how you add to calculate the total.

a. $3\frac{4}{5} + 2\frac{2}{3}$	Same denominators	Show addition

b. $2\frac{3}{4} + 1\frac{5}{6}$	Same denominators	Show addition

FUNNY SPORT

Why should you not tell jokes when you ice-skate?

★ Figure out each of these and write the product. Then write each letter above its matching product below. Some letters appear more than once.

6 × 35 = ___ **h**	15 × 7 = ___ **e**	45 × 6 = ___ **a**
35 × 8 = ___ **t**	8 × 15 = ___ **r**	15 × 12 = ___ **i**
35 × 12 = ___ **u**	25 × 16 = ___ **g**	24 × 15 = ___ **k**
45 × 16 = ___ **c**	35 × 16 = ___ **m**	16 × 15 = ___ **p**

___ ___ ___   ___ ___ ___
280 210 105   180 720 105

___ ___ ___ ___ ___   ___ ___ ___ ___ ___   ___ ___
560 180 400 210 280   720 120 270 720 360   420 240

Write these products.

15 × 28 = ___   18 × 15 = ___   14 × 15 = ___

12 × 25 = ___   35 × 14 = ___   45 × 12 = ___

18 × 35 = ___   8 × 45 = ___   15 × 22 = ___

1. Write an expression that uses common fractions to break each number into parts.

a. 4.612 $4 + \frac{6}{10} +$ ___

b. 5.061   ___

c. 2.402   ___

2. Solve these problems. Show your thinking.

a. The dimensions of a storage room are 8 ft × 9 ft × 5 ft. What is the volume of the room?

_____ ft³

b. The back of a tip truck measures 12 ft × 6 ft × 3 ft. What is its volume?

_____ ft³

3. Write the missing numbers. Then write the answers.

a. 18 + 5 = ___
   12 + 6 + 5
   12 + 11 = ___

b. ___ − 5 = ___
   21 − 7 − 5
   21 − ___ = ___

c. ___ × 4 = ___
   6 × 5 × 4
   6 × ___ = ___

d. ___ × 5 = ___
   7 × 3 × 5
   7 × ___ = ___

e. ___ ÷ 3 = ___
   54 ÷ 9 ÷ 3
   54 ÷ ___ = ___

f. ___ ÷ 3 = ___
   18 ÷ 6 ÷ 3
   18 ÷ ___ = ___

ORIGO Stepping Stones 5 4.12

1. Draw lines to connect matching values.

| 1,000,000 | 10,000 | 1,000 | 100,000 | 10 | 100 |

| 10^3 | 10^4 | 10^5 | 10^6 | 10^2 | 10^1 |

2. For each of these, write the denominator that is common to both fractions and complete the equivalent fractions. Then write the totals.

a. $\frac{1}{3} + \frac{2}{5} = \frac{\ }{\ }$

☐/☐ + ☐/☐ = ☐/☐

b. $\frac{1}{5} + \frac{3}{4} = \frac{\ }{\ }$

☐/☐ + ☐/☐ = ☐/☐

c. $\frac{1}{4} + \frac{4}{6} = \frac{\ }{\ }$

☐/☐ + ☐/☐ = ☐/☐

d. $\frac{3}{5} + \frac{1}{3} = \frac{\ }{\ }$

☐/☐ + ☐/☐ = ☐/☐

e. $\frac{4}{5} + \frac{5}{6} = \frac{\ }{\ }$

☐/☐ + ☐/☐ = ☐/☐

f. $\frac{1}{2} + \frac{5}{9} = \frac{\ }{\ }$

☐/☐ + ☐/☐ = ☐/☐

3. a. Write these decimals in order from **least** to **greatest**. Use the number line to help you.

| 0.720 | 0.702 | 0.027 | 0.072 |

0 ————————————————— 1

b. In each group, loop the greatest number.

| 0.602 | 0.062 | 0.620 |

| 0.470 | 0.407 | 0.047 |

HOW LONG?

How long should an adult giraffe's legs be?

★ Figure out each of these and write the product. Then write each letter above its matching product at the bottom of the page. Some letters appear more than once.

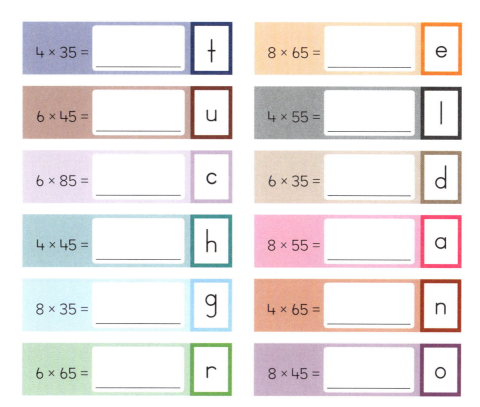

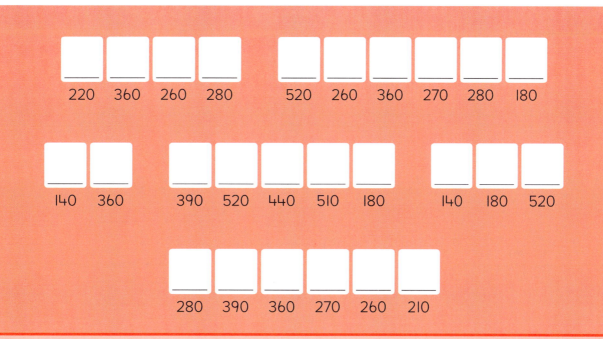

1. Rewrite both fractions so the denominators are the same.

a. $\frac{3}{4} = \dfrac{}{}$ $\frac{1}{5} = \dfrac{}{}$

b. $\frac{2}{3} = \dfrac{}{}$ $\frac{1}{2} = \dfrac{}{}$

c. $\frac{1}{3} = \dfrac{}{}$ $\frac{3}{5} = \dfrac{}{}$

d. $\frac{2}{5} = \dfrac{}{}$ $\frac{5}{8} = \dfrac{}{}$

e. $\frac{6}{4} = \dfrac{}{}$ $\frac{2}{6} = \dfrac{}{}$

f. $\frac{3}{5} = \dfrac{}{}$ $\frac{8}{6} = \dfrac{}{}$

2. Figure out the totals using improper fractions and then mixed numbers.

a. $2\frac{4}{12} + 3\frac{2}{6}$

Use improper fractions	Use mixed numbers

b. $1\frac{1}{4} + 3\frac{5}{8}$

Use improper fractions	Use mixed numbers

c. $2\frac{3}{8} + 3\frac{1}{2}$

Use improper fractions	Use mixed numbers

3. Read the number on the expander. Then round each number to the nearest **whole number**, **tenth**, and **hundredth**.

a. 4.193

b. 2.746

c. 6.904

	Nearest whole number	Nearest tenth	Nearest hundredth
a.			
b.			
c.			

1. Draw a line to show the location of each decimal fraction. Be as accurate as possible.

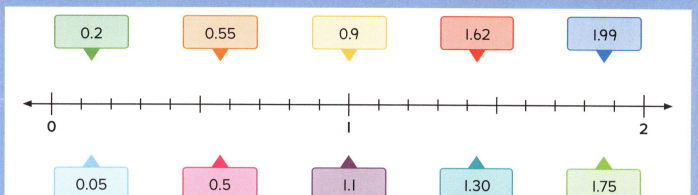

2. Figure out the totals using improper fractions and then mixed numbers.

a.	Use improper fractions	Use mixed numbers
$3\frac{2}{6} + 1\frac{1}{3}$		

b.	Use improper fractions	Use mixed numbers
$2\frac{3}{6} + 4\frac{1}{4}$		

3. For each decimal below, draw an arrow to show where you think it is on the number line. Then write the whole number that is closest.

a. 3.59   b. 3.95   c. 4.25   d. 5.02   e. 5.75   f. 6.40

g. 16.10   h. 16.51   i. 17.6   j. 18.42   k. 19.33   l. 20.09

BELIEVE IT!

★ Figure out each of these and draw a straight line to the correct answer. The line will pass through a number and a letter. Write each letter above its matching number at the bottom of the page. Some letters appear more than once.

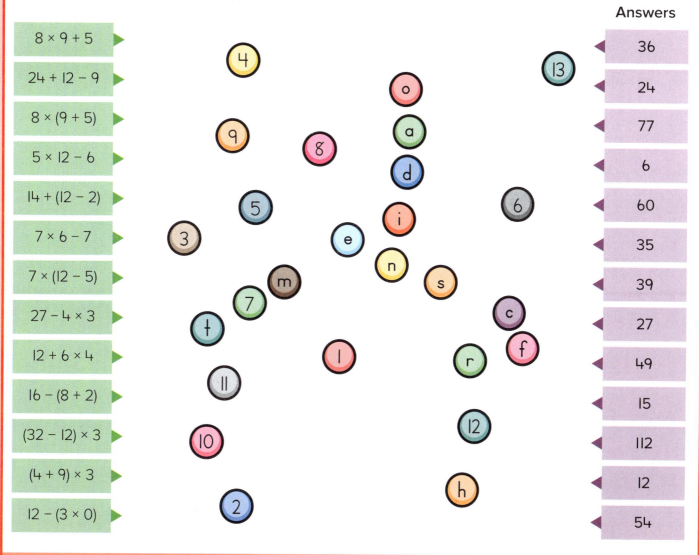

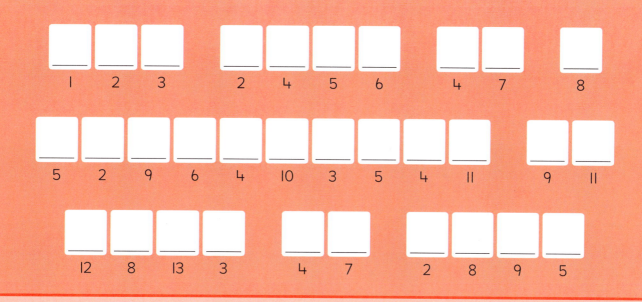

1. Complete these.

a.
```
    7  5  2
 ×         6
 ─────────────
```

b.
```
    4  0  8
 ×         7
 ─────────────
```

c.
```
 1  6  1  5
 ×         8
 ─────────────
```

d.
```
 2  6  5  4
 ×         9
 ─────────────
```

2. Rewrite the mixed numbers so the fractions have the same denominator. Then show how you add to calculate the total.

a. $2\frac{3}{4} + 3\frac{2}{5}$

Same denominators	Show addition

b. $3\frac{4}{6} + 4\frac{3}{4}$

Same denominators	Show addition

3. Figure out each total. Draw jumps to show your thinking.

a. 7.4 + 2.55 = _____

b. 3.12 + 4.2 = _____

c. 5.09 + 1.4 = _____

ORIGO Stepping Stones 5 · 5.8

1. Complete these.

a.
```
    4 6 2
  ×   4 7
  _____
```

b.
```
    7 0 6
  × 5 2 3
  _____
```

c.
```
    3 8 9 0
  ×   2 7 4
  _____
```

2. Color each rectangle to show how you could split it into two parts to figure out the area. Complete each equation.

a.
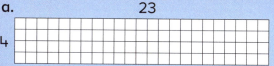
23
4

4 × (_____ + ____) = ____ × _____ + ____ × _____

b.

24
8

8 × (_____ + ____) = ____ × _____ + ____ × _____

3. Figure out the total. Show your thinking.

a.

$ _____

b. $3.70

$ _____

c. $5.19 $3.55

$ _____

d. $2.45

$ _____

WORM WONDERS

How can you tell which end of a worm is the head?

★ Figure out each of these and write the answer. Then write each letter above its matching answer at the bottom of the page.

550 ÷ 5 = ___ t
690 ÷ 3 = ___ i
480 ÷ 4 = ___ e
720 ÷ 3 = ___ a
660 ÷ 5 = ___ h
720 ÷ 4 = ___ k
420 ÷ 3 = ___ l
780 ÷ 3 = ___ n
480 ÷ 5 = ___ w
520 ÷ 4 = ___ c
450 ÷ 3 = ___ e
810 ÷ 5 = ___ c
480 ÷ 3 = ___ h
850 ÷ 5 = ___ d
520 ÷ 5 = ___ m
720 ÷ 5 = ___ l
240 ÷ 3 = ___ m
570 ÷ 3 = ___ w
630 ÷ 5 = ___ l
590 ÷ 5 = ___ t

___ i ___ ___ e i t i ___
118 130 180 144 260

t ___ e ___ i ___ d e
132 80 170 140

___ n d ___ a ___ c ___ ___ h ___ ___ h
240 190 110 160 96 230 162

___ n d s ___ i ___ ___ s
150 104 126 120

ORIGO Stepping Stones 5 • 5.11

44

1. Complete each table to show the dimensions of three different prisms that have the same volume.

a. Volume = 48 in³

Length	Width	Height

b. Volume = 60 in³

Length	Width	Height

2. In each equation, draw parentheses if they are needed to make it true.

a. 16 − 8 × 3 = 24

b. 4 = 2 × 16 ÷ 8

c. 7 + 5 × 2 ÷ 12 = 2

d. 42 + 18 + 15 = 75

e. 35 ÷ 7 + 4 × 6 = 29

f. 200 = 100 ÷ 2 × 4

3. Calculate the total mass of each pair of packages.

A: 3.6 kg
B: 1.48 kg
C: 9.08 kg
D: 6.74 kg
E: 10.36 kg

a. B + D

b. D + E

c. A + C

d. A + B

e. C + E

f. B + C

1. Complete the missing parts.

a.

three and _____

b.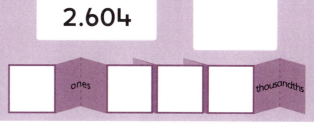

_____

2. a. Write these decimals in order from **least** to **greatest**. Use the number line to help you.

0.45   0.045   0.540   0.504

b. Write these decimals in order from **greatest** to **least**.

0.1   0.129   0.219   0.902   1.29   0.2

_____ _____ _____ _____ _____ _____

3. Complete each equation. Use the number line to help rewrite one or both numbers.

a. $2\frac{2}{5} - \frac{4}{5} = \boxed{}$

b. $2\frac{3}{6} - \frac{10}{6} = \boxed{}$

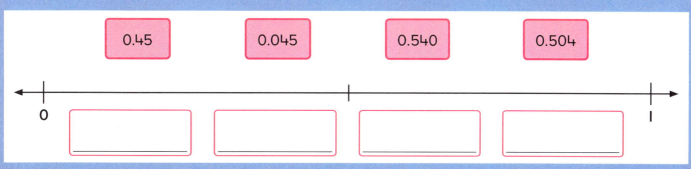

DISTANCE

What is the difference between *here* and *there*?

★ Figure out each of these and write the product. Then find the product in the puzzle below and shade the matching letter. The remaining letters will spell the answer.

3 × 260 =	140 × 5 =	170 × 3 =
4 × 180 =	3 × 160 =	190 × 5 =
280 × 3 =	240 × 4 =	3 × 310 =
250 × 3 =	3 × 130 =	160 × 4 =
180 × 5 =	190 × 4 =	5 × 130 =
140 × 3 =	3 × 290 =	230 × 4 =
3 × 180 =		

1. Complete these.

a.
```
      6 3
×     2 7
_____
```

b.
```
      4 2
×     3 6
_____
```

c.
```
      2 9
×     5 8
_____
```

d.
```
      8 2
×     6 4
_____
```

2. Calculate these perimeters. Record the steps you use.

a.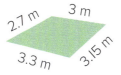

_____ m

b.

_____ m

c.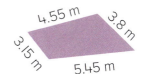

_____ m

3. Complete each equation. Rewrite the fractions so the denominators are the same. Use the number line to help.

a. $\dfrac{3}{5} - \dfrac{4}{10} = \boxed{}$

b. $\dfrac{17}{20} - \dfrac{5}{10} = \boxed{}$

c. $\dfrac{8}{10} - \dfrac{3}{5} = \boxed{}$

Number line: 0, $\dfrac{1}{20}$, $\dfrac{1}{10}$, $\dfrac{1}{5}$, ... 1

1. Complete these.

a.
```
    2 6 7
×     3 5
─────────
```

b.
```
    6 8 3
×   3 1 4
─────────
```

c.
```
  2 0 7 9
×   1 4 5
─────────
```

d.
```
    5 2 3
×     4 5
─────────
```

e.
```
    7 1 9
×   2 2 3
─────────
```

f.
```
  2 7 0 9
×   1 5 4
─────────
```

2. Describe this polygon. Be sure to mention its angles and sides.

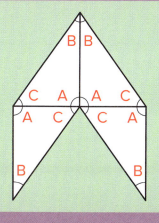

_____
_____
_____
_____
_____

3. Complete each equation. Rewrite **one** or **both** fractions so the denominators are the same.

a. $\dfrac{3}{4} - \dfrac{1}{3} = \boxed{}$

b. $\dfrac{4}{5} - \dfrac{4}{6} = \boxed{}$

c. $\dfrac{4}{5} - \dfrac{3}{4} = \boxed{}$

RAIN, RAIN, GO AWAY

Three tall men were standing under one umbrella and none of them got wet. How could that be?

★ Figure out each of these and write the total. Then write each letter above its matching total at the bottom of the page. Some letters appear more than once.

$3.28 + $4.65 = $ _____ **r**	$2.38 + $6.47 = $ _____ **u**
$2.46 + $3.57 = $ _____ **n**	$5.28 + $1.59 = $ _____ **e**
$5.54 + $1.18 = $ _____ **y**	$3.45 + $3.26 = $ _____ **a**
$1.46 + $2.39 = $ _____ **h**	$2.46 + $5.17 = $ _____ **l**
$4.37 + $2.56 = $ _____ **d**	$3.56 + $3.18 = $ _____ **b**
$5.44 + $3.38 = $ _____ **g**	$4.32 + $2.49 = $ _____ **i**
$1.53 + $2.38 = $ _____ **t**	$7.58 + $1.28 = $ _____ **w**
$3.46 + $2.28 = $ _____ **s**	

___ ___ ___ ___   ___ ___ ___ ___
$3.91 $3.85 $6.87 $6.72 $8.86 $6.87 $7.93 $6.87

___ ___ ___ ___ ___ ___   ___
$6.81 $6.03 $5.74 $6.81 $6.93 $6.87 $6.71

___ ___ ___ ___ ___ ___ ___
$6.74 $8.85 $6.81 $7.63 $6.93 $6.81 $6.03 $8.82

1. Complete this table. You can use base-10 ones blocks to help.

Number of cubes in base	Number of layers	Total number of cubes
7	1	
	3	
	5	
	7	
	9	

2. a. Draw a ✔ on each shape that is a parallelogram.

b. Write how you know.

_____
_____

3. Figure out the differences using improper fractions and then mixed numbers.

a. $3\frac{7}{8} - 1\frac{3}{4}$

Use improper fractions	Use mixed numbers

b. $4\frac{4}{6} - 2\frac{1}{3}$

Use improper fractions	Use mixed numbers

1. Solve these problems. Show your thinking.

   a. Three bags of fruit weighed $2\frac{3}{4}$ lb, $3\frac{1}{2}$ lb, and $4\frac{1}{4}$ lb. What was the total weight of all the fruit?

   _____ lb

   b. Three lengths of pipe measured $4\frac{2}{3}$ yd, $5\frac{3}{4}$ yd, and $3\frac{1}{2}$ yd. What was the total length of pipe?

   _____ yd

2. Write **P** inside the parallelograms.

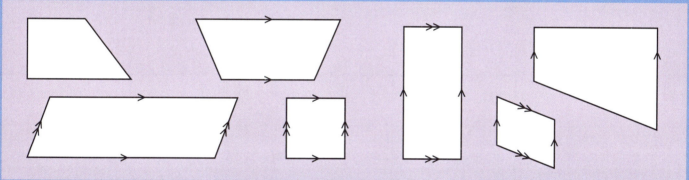

3. Figure out the differences using improper fractions and then mixed numbers.

   a. $3\frac{2}{3} - 1\frac{1}{5}$

Use improper fractions	Use mixed numbers

   b. $2\frac{3}{4} - 1\frac{2}{3}$

Use improper fractions	Use mixed numbers

FLIP OUT!

★ Figure out each of these and draw a straight line to the correct total. The line will pass through a letter. Write each letter above its matching total at the bottom of the page. Some letters appear more than once.

Expression		Totals
2.25 + 4.75 + 5.3		18.2
1.05 + 3.65 + 4.4		13.1
7.95 + 2.05 + 3.2		8.2
2.33 + 4.3 + 6.47		11.3
3.41 + 5.6 + 2.29		12.3
1.38 + 2.8 + 4.12		8.3
4.36 + 2.24 + 3.2		13.65
5.54 + 2.16 + 5.2		13.2
8.4 + 2.65 + 7.15		9.53
3.3 + 3.72 + 1.18		9.1
4.4 + 5.6 + 3.65		9.8
1.7 + 4.53 + 3.3		12.9

Letters in the middle: p, l, d, n, s, h, e, o, t, i, y, w

___ ___ ___ ___ ___ ___ ___ ___
11.3 9.1 13.1 12.3 18.2 9.8 13.2 8.3

___ ___ ___ ___ ___   ___ ___ ___ ___
8.3 13.1 8.2 8.2 12.3   12.9 9.8 13.65 18.2

___ ___ ___   ___ ___ ___   ___ ___ ___ ___
9.1 13.2 8.2   8.2 9.53 8.2   9.1 12.3 8.2 13.2

1. Read the expression. Complete all of the equations.

a. 3 × 27

3 × (20 + 7) = 3 × _____ + 3 × _____

3 × (25 + 2) = 3 × _____ + 3 × _____

3 × 27 = _____

b. 4 × 48

4 × (40 + 8) = 4 × _____ + _____ × _____

4 × (45 + 3) = _____ × _____ + _____ × _____

4 × 48 = _____

c. 6 × 36

6 × (30 + 6) = 6 × _____ + _____ × _____

6 × (35 + 1) = _____ × _____ + _____ × _____

6 × 36 = _____

d. 4 × 38

4 × (30 + 8) = _____ × _____ + _____ × _____

4 × (35 + 3) = _____ × _____ + _____ × _____

4 × 38 = _____

2. Draw and label one equilateral triangle, one isosceles triangle, and one scalene triangle.

3. Rewrite each mixed number so that the fractions have the same denominator. Then subtract to figure out the difference.

	Same denominators	Show subtraction
a. $3\frac{1}{4} - 1\frac{3}{6}$		
b. $4\frac{2}{5} - 2\frac{2}{3}$		

ORIGO Stepping Stones 5 · 6.12

1. Complete these.

a. $2 \times \frac{4}{5} =$ ☐

b. $3 \times \frac{3}{4} =$ ☐

c. $4 \times \frac{6}{8} =$ ☐

d. $2 \times \frac{2}{3} =$ ☐

e. ☐ $= 5 \times \frac{2}{6}$

f. ☐ $= 4 \times \frac{3}{5}$

g. $2 \times \frac{7}{8} =$ ☐

h. ☐ $= 3 \times \frac{4}{12}$

2. Complete each equation. Show your thinking.

a. $3\frac{2}{5} - \frac{4}{5} =$ ☐

b. $\frac{2}{3} - \frac{1}{6} =$ ☐

c. $\frac{4}{5} - \frac{2}{4} =$ ☐

3. Draw jumps on the number line to figure out each difference.

a. $8.7 - 3.4 =$ ___

b. $6.4 - 3.2 =$ ___

c. $7.6 - 5.3 =$ ___

QUIZ QUESTION

What are red velvet swords, firemouths, and tiger barbs?

★ Figure out each of these and rule a line to the correct answer. The line will pass through a number and a letter. Write each letter above its matching number at the bottom of the page.

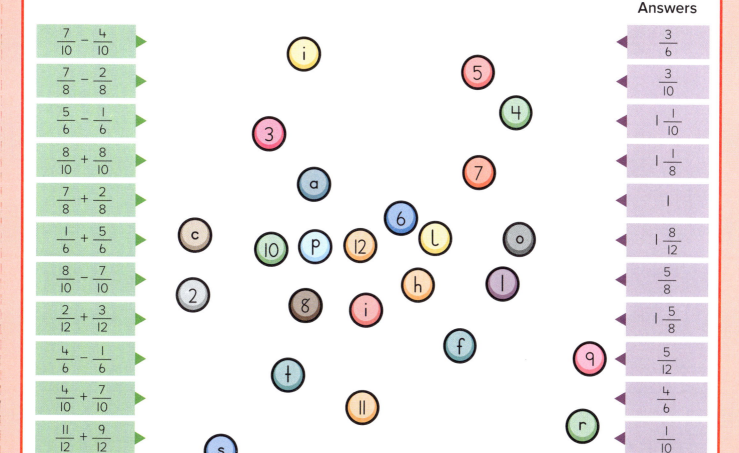

1. Rewrite each equation so the denominators are the same. Then write the total.

a. $\frac{1}{3} + \frac{2}{6} = \frac{}{}$

b. $\frac{5}{8} + \frac{1}{4} = \frac{}{}$

c. $\frac{2}{5} + \frac{3}{15} = \frac{}{}$

d. $\frac{2}{4} + \frac{2}{5} = \frac{}{}$

e. $\frac{3}{5} + \frac{1}{3} = \frac{}{}$

f. $\frac{3}{12} + \frac{4}{8} = \frac{}{}$

2. Complete each equation. Show your thinking.

a. $2\frac{2}{3} - 1\frac{1}{6} = \square$

b. $3\frac{3}{4} - 1\frac{2}{6} = \square$

c. $4\frac{1}{4} - 1\frac{1}{3} = \square$

3. Figure out the amount that is left in the wallet after each purchase.

a. $16.70 $3.25

$ _____

b. $18.57 $7.25

$ _____

ORIGO Stepping Stones 5 · 7.4

1. Write an expression to show how to solve each problem. You do not need to figure out the answer.

   a. Mary has $20. How much change will she receive if she buys 6 books at $2.50 each?

   b. How much money will Damon need to buy 3 t-shirts at $12 each and a pair of shorts for $24?

2. Convert these measurements.

   a. 18 inches = ____ ft ____ in

   b. 15 inches = ____ ft ____ in

   c. 23 inches = ____ ft ____ in

   d. 32 inches = ____ ft ____ in

   e. 2.5 feet = ____ in

   f. 3.75 feet = ____ in

3. Use a written method to figure out the difference between each pair of weights.

   a. 3.1 kg 6.48 kg

   _____ kg

   b. 8.59 kg 4.05 kg

   _____ kg

   c. 9.06 kg 12.3 kg

   _____ kg

   d. 4.5 kg 15.81 kg

   _____ kg

ANY IDEA?

What was more useful than the invention of the first telephone?

★ Figure out each of these and write the total. Then write each letter above its matching total at the bottom of the page.

55¢ + $3.95 + $4.40 = $____	e	$7.25 + 35¢ + 20¢ = $____	e
$4.05 + $1.60 + 30¢ = $____	t	35¢ + $2.50 + $4.05 = $____	p
$2.50 + 25¢ + $3.15 = $____	l	$3.45 + $1.40 + 10¢ = $____	t
$2.15 + $3.20 + $1.30 = $____	h	35¢ + $1.60 + $2.00 = $____	n
$4.35 + $2.20 + $3.20 = $____	c	$2.30 + $1.15 + $1.45 = $____	o
$7.45 + $1.05 + $1.35 = $____	o	$2.10 + $3.55 + $1.15 = $____	e
$5.10 + $1.35 + $2.25 = $____	n	$3.15 + $2.20 + $3.45 = $____	d
$4.25 + $1.40 + $2.25 = $____	e	$3.20 + $2.30 + $3.35 = $____	s
$3.05 + $2.30 + $1.15 = $____	e	$1.65 + $4.05 + $2.00 = $____	h

____ ____ ____   ____ ____ ____ ____ ____ ____
$4.95 $6.65 $8.90 $8.85 $7.90 $9.75 $4.90 $8.70 $8.80

____ ____ ____ ____ ____ ____ ____ ____ ____
$5.95 $6.50 $5.90 $7.80 $6.90 $7.70 $9.85 $3.95 $6.80

1. Write **<**, **>**, or **=** to make each statement true.

a. 0.525 ◯ 0.552   b. 0.07 ◯ 0.7   c. 0.6 ◯ 0.599

d. 0.1 ◯ 0.100   e. 0.412 ◯ 0.421   f. 0.39 ◯ 0.309

g. 0.780 ◯ 0.87   h. 0.11 ◯ 1.1   i. 0.16 ◯ 0.160

2. Convert the measurements.

a. 4 yd = _____ ft   b. 8 yd = _____ ft   c. $6\frac{1}{2}$ yd = _____ ft

d. 11 yd = _____ ft   e. 15 yd = _____ ft   f. $18\frac{1}{2}$ yd = _____ ft

g. _____ yd = 12 ft   h. _____ yd = 9 ft   i. _____ yd = 27 ft

j. _____ yd = 39 ft   k. _____ yd = 21 ft   l. _____ yd = 45 ft

3. Draw jumps on the number line to figure out each difference.

a. 8.4 − 6.8 = _____

b. 5.2 − 2.7 = _____

c. 7.6 − 3.9 = _____

1. For each number, draw an arrow to show where you think it is on the number line. Then round it to the nearest tenth and write the tenth.

a. 0.21 b. 0.435 c. 0.97 d. 1.08 e. 1.49 f. 1.905

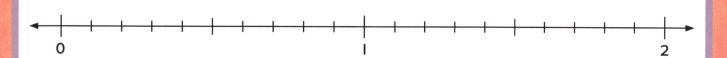

2. Solve each problem. Write the answer two ways. Show your thinking.

a. On the ninth green Kimie had to sink an 18 foot putt. The ball stopped $4\frac{1}{2}$ ft past the hole. What is the length of her next putt?

_____ yards   _____ inches

b. Saturday's first high tide was $2\frac{1}{2}$ yards. Its second high tide was 6 inches higher. What was the height of Saturday's second high tide?

_____ feet   _____ inches

3. Figure out each difference. Show your thinking.

a. 24.6 − 19.8 = _____

b. 18.05 − 7.12 = _____

c. 15.3 − 12.85 = _____

d. 36.25 − 9.8 = _____

e. 41.37 − 8.50 = _____

f. 21.5 − 0.8 = _____

WEIRD AND WONDERFUL

What can be heard but not seen, and only speaks when spoken to?

★ Figure out each of these and write the total. Find each total in the grid below and cross out the letter above. Then write the remaining letters at the bottom of the page.

14.87 + 12.6 =	4.5 + 13.72 =	14.38 + 6.02 =
12.53 + 11.64 =	13.56 + 11.7 =	5.6 + 14.85 =
13.56 + 3.05 =	13.47 + 14.72 =	12.75 + 12.6 =
7.4 + 13.42 =	11.74 + 4.5 =	10.36 + 18.85 =
11.64 + 15.7 =	8.3 + 16.38 =	13.47 + 5.05 =
11.56 + 13.86 =	16.84 + 3.5 =	5.6 + 14.54 =
9.64 + 3.5 =		

✻

D	A	I	S	Y	C	H	A	I	N
20.45	24.35	20.34	27.47	16.24	25.42	25.35	24.17	24.68	20.58

M	E	D	A	L	C	A	T	C	H
18.22	16.51	29.21	16.61	20.14	19.59	25.26	20.82	18.52	12.69

H	O	O	P	S
27.34	13.14	15.79	20.4	28.19

Write the letters in order from the ✻ to the bottom-right corner.

1. Figure out each total. Show your thinking.

a. $3.25 $5.40

$ _____

b. $4.75 $3.50

$ _____

c. $6.82 $2.45

$ _____

2. Use this line plot to answer the questions below.

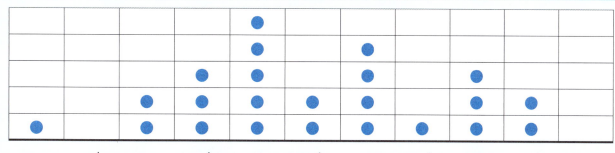

Class Reactions (inches)

a. How many results were recorded? _____

b. What result was recorded most frequently? _____ inches

c. What was the shortest result? _____ inches

d. How many students recorded the longest result? _____

e. What was the difference between the length of the longest and shortest results? _____ inches

3. Calculate the difference in mass between these sacks of grain. Record the steps you use.

a. 15.35 kg   7.6 kg

_____ kg

b. 4.25 kg   12.1 kg

_____ kg

c. 3.85 kg   7.12 kg

_____ kg

ORIGO Stepping Stones 5 7.12

1. Complete the missing numbers to calculate each product.

a. $8 \times 2\frac{3}{10}$

$(8 \times \underline{}) + (8 \times \underline{})$

$\underline{} + \underline{} = \underline{}$

b. $5 \times 4\frac{3}{8}$

$(5 \times \underline{}) + (5 \times \underline{})$

$\underline{} + \underline{} = \underline{}$

2. Figure out the amount left in the wallet after each purchase.

a. $9.45 $3.05

$ _____

b. $12.43 $5.89

$ _____

c. $6.58 $1.25

$ _____

3. Complete each of these. Show your thinking.

a. $348 \div 4 = \underline{}$

b. $435 \div 3 = \underline{}$

c. $762 \div 6 = \underline{}$

d. $672 \div 8 = \underline{}$

e. $595 \div 5 = \underline{}$

f. $846 \div 6 = \underline{}$

DOUBLE VISION

Why should a golfer wear two pairs of shorts?

★ Figure out each of these and write the product. Then write each letter above its matching product at the bottom of the page.

27 × 33 = ____ i	42 × 21 = ____ t
34 × 15 = ____ n	45 × 13 = ____ a
24 × 32 = ____ e	16 × 23 = ____ e
28 × 12 = ____ t	18 × 41 = ____ e
25 × 48 = ____ h	22 × 37 = ____ e
33 × 14 = ____ l	31 × 26 = ____ h
35 × 25 = ____ i	19 × 21 = ____ n
15 × 41 = ____ m	13 × 26 = ____ o
48 × 12 = ____ g	21 × 28 = ____ h
22 × 44 = ____ g	17 × 26 = ____ o

1,200 768   615 891 968 806 336   576 814 882

585   588 338 462 368   875 510   442 399 738

1. Rewrite both fractions so they have the same denominator. Then write the total.

a. $\frac{1}{4} + \frac{3}{5} = \frac{\ \ }{\ \ }$

b. $\frac{2}{3} + \frac{1}{5} = \frac{\ \ }{\ \ }$

c. $\frac{3}{4} + \frac{1}{6} = \frac{\ \ }{\ \ }$

2. Draw jumps on the number line to figure out each difference.

a. $8.40 - 5.75 =$ _____

b. $6.10 - 3.95 =$ _____

c. $7.60 - 4.85 =$ _____

3. Complete these. Show your thinking.

a. $7{,}524 \div 4 =$ _____

b. $7{,}449 \div 3 =$ _____

c. $9{,}615 \div 5 =$ _____

ORIGO Stepping Stones 5 · 8.4

1. Rewrite both numbers so the fractions have the same denominator. Then write the total.

a. $2\frac{1}{3} + 3\frac{2}{4} =$ ☐

b. $1\frac{1}{3} + 3\frac{1}{2} =$ ☐

c. $4\frac{2}{6} + 2\frac{1}{4} =$ ☐

2. Mark the three corners of each rectangular-shaped building on the coordinate plane below. Write the coordinates of the 4th corner. Then shade the buildings.

Bank
(2, 5)
(5, 5)
(5, 1)
(___ , ___)

Mall
(9, 11)
(14, 11)
(9, 7)
(___ , ___)

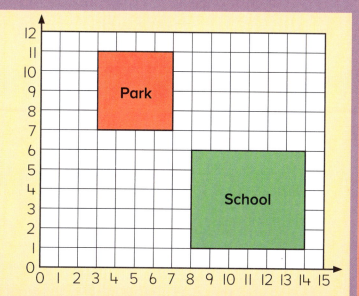

3. Figure out the amount in each share.

a. $672 \div 6 =$ _____

100
100
100
100
100
100

b. $805 \div 7 =$ _____

c. $917 \div 7 =$ _____

ORIGO Stepping Stones 5 · 8.6

67

DID YOU KNOW?

★ To find out, figure out each of these and write the difference. Then write each letter above its matching difference at the bottom of the page.

35 − 7.8 =	l	25 − 6.7 =	t
27 − 8.8 =	s	23 − 6.5 =	m
28 − 4.8 =	l	21 − 5.8 =	v
42 − 6.9 =	e	34 − 8.8 =	h
41 − 6.7 =	a	33 − 4.6 =	y
43 − 8.5 =	d	22 − 6.9 =	e
18 − 4.7 =	e	26 − 7.9 =	e
40 − 5.9 =	h	35 − 6.7 =	e
36 − 8.9 =	e	34 − 2.8 =	i
42 − 7.8 =	r	23 − 5.6 =	a
27 − 9.8 =	c	51 − 7.5 =	s

17.2 34.3 16.5 27.1 27.2 18.2   34.1 17.4 15.2 15.1

18.3 25.2 34.2 13.3 28.3   35.1 28.4 18.1 23.2 31.2 34.5 43.5

1. Draw an arrow to show the exact position of each decimal fraction on the number line. Then write the nearest **hundredth**.

   a. 1.291   b. 1.296   c. 1.299   d. 1.303   e. 1.308

2. Look at this growing pattern.

   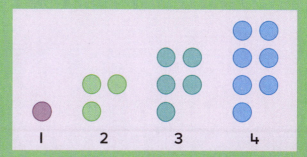

   a. Complete this table. Continue the pattern.

Picture number	1	2	3	4	5	6
Number of dots	1	3				

   b. How did you figure out the numbers to write in the second row?

   _____

3. Use the standard division algorithm to calculate each quotient.

   a. 7) 84   b. 8) 96   c. 5) 95   d. 3) 72

   (each shown in T | O place value format)

1. Complete the missing numbers to calculate each product.

a. $4 \times 2\frac{5}{6}$

(4 × ___) + (4 × ___)

___ + ___ = ___

b. $3 \times 6\frac{3}{4}$

(3 × ___) + (3 × ___)

___ + ___ = ___

2. a. Complete this table to show the number of cards needed, if each player gets 4 cards.

Number of players	1	2	3	4	5
Number of cards	4				

b. Write the data from the table above as ordered pairs.

(___ , ___) (___ , ___) (___ , ___)

(___ , ___) (___ , ___)

c. Plot the ordered pairs on the coordinate plane.

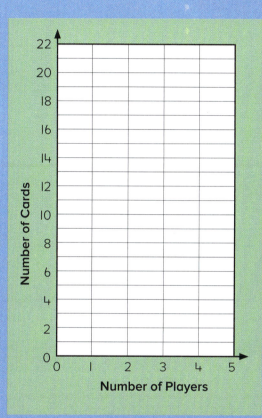

3. Complete these calculations using the standard division algorithm.

a. H T O

3) 7 2 6

b. H T O

4) 9 7 6

c. H T O

5) 8 5 5

ORIGO Stepping Stones 5 · 8.10   70

PICK THE POLITICIAN

Who was the first President to live in the White House?

★ Figure out each of these and write the answer. Then find each answer in the puzzle below and shade the matching letter. The remaining letters will spell the answer.

$1\frac{5}{8} + \frac{5}{8} =$ ___	$1\frac{3}{8} - \frac{5}{8} =$ ___	$1\frac{3}{5} + 1\frac{3}{5} =$ ___
$1\frac{1}{6} - \frac{4}{6} =$ ___	$\frac{7}{12} + \frac{11}{12} =$ ___	$3 - \frac{4}{5} =$ ___
$\frac{7}{10} + 1\frac{1}{10} =$ ___	$1\frac{1}{10} - \frac{7}{10} =$ ___	$1\frac{1}{6} - \frac{5}{6} =$ ___
$\frac{3}{8} + 2\frac{7}{8} =$ ___	$1\frac{7}{12} - \frac{11}{12} =$ ___	$\frac{4}{5} + 1\frac{3}{5} =$ ___
$\frac{1}{6} + 1\frac{5}{6} =$ ___	$2\frac{1}{8} - 1\frac{4}{8} =$ ___	$1\frac{9}{12} + 1\frac{9}{12} =$ ___
$2\frac{3}{6} - 1\frac{5}{6} =$ ___	$2\frac{3}{8} - 1\frac{2}{8} =$ ___	$\frac{4}{5} + 1\frac{4}{5} =$ ___
$1\frac{3}{8} - \frac{4}{8} =$ ___	$3\frac{4}{5} - 3\frac{2}{5} =$ ___	

ORIGO Stepping Stones 5 • 8.11   71

1. Complete each equation. Show your thinking.

a. $3\frac{3}{4} - 1\frac{3}{8} = \boxed{}$

b. $4\frac{1}{2} - 2\frac{2}{5} = \boxed{}$

c. $5\frac{1}{4} - 2\frac{1}{5} = \boxed{}$

2. a. These items are on sale. Each item now costs $5 less. Complete the table.

	Games	CDs	Puzzles	Books
Original price	$12	$7	$6	$14
Sale price				

b. Write ordered pairs to match the data in the table.

Games (___ , ___)

CDs (___ , ___)

Puzzles (___ , ___)

Books (___ , ___)

c. Plot the ordered pairs on the coordinate plane.

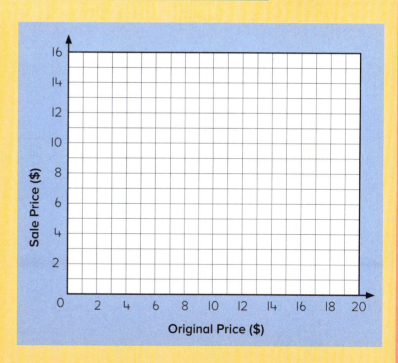

3. Convert each length. Then write it as a fraction of a meter.

a. 0.6 m = _____ cm = $\frac{6}{10}$ m

b. 0.25 m = _____ cm = _____ m

c. _____ m = 210 cm = _____ m

d. _____ m = 90 cm = _____ m

1. Draw an arrow to connect each decimal fraction to its approximate position on the number line. Be as accurate as possible.

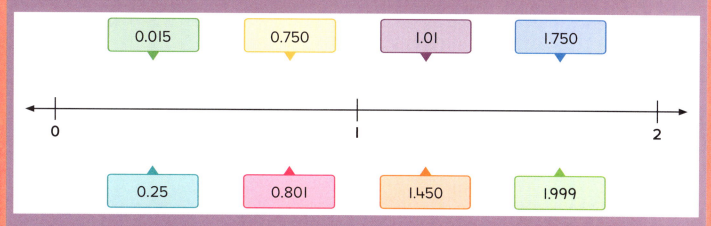

2. Rewrite each equation using the division bracket.

a. $72 \div 3 = 24$

T O

b. $21 = 84 \div 4$

T O

c. $107 = 535 \div 5$

H T O

d. $408 \div 4 = 102$

H T O

3. Calculate each product. Show your thinking.

a. $\frac{2}{3} \times 4$

b. $\frac{2}{5} \times 3$

c. $6 \times 4\frac{1}{3}$

d. $5 \times 3\frac{2}{6}$

e. $5 \times 2\frac{3}{4}$

f. $4 \times 5\frac{1}{4}$

ORIGO Stepping Stones 5 · 9.2

SUPER MAN?

Throw me off the tallest building and I will not break, but put me in the ocean and I will. What am I?

★ Figure out each of these and write the quotient. Then find each quotient in the puzzle below and shade the matching letter. The remaining letters will spell the answer.

524 ÷ 4 = _____   423 ÷ 3 = _____   695 ÷ 5 = _____

345 ÷ 5 = _____   656 ÷ 4 = _____   276 ÷ 3 = _____

528 ÷ 3 = _____   485 ÷ 5 = _____   472 ÷ 4 = _____

396 ÷ 4 = _____   384 ÷ 3 = _____   565 ÷ 5 = _____

275 ÷ 5 = _____   264 ÷ 4 = _____   441 ÷ 3 = _____

243 ÷ 3 = _____   395 ÷ 5 = _____   192 ÷ 4 = _____

516 ÷ 4 = _____   534 ÷ 3 = _____   785 ÷ 5 = _____

1. Complete these.

a.
```
    5 1 6
  ×     7
  ———————
```

b.
```
    3 8 7
  ×     6
  ———————
```

c.
```
  1 5 0 3
  ×     8
  ———————
```

d.
```
      3 7 2
  ×   1 6 5
  ———————
```

e.
```
    4 8 3
  ×     9
  ———————
```

f.
```
    6 2 3
  ×     4
  ———————
```

g.
```
    2 7 1 2
  ×       7
  ———————
```

2. Complete the missing parts to make these true.

a. 1.4 m = ☐ cm = $1\frac{4}{10}$ m

b. ☐ m = 150 cm = ☐ m

c. ☐ m = ☐ cm = $2\frac{75}{100}$ m

d. ☐ m = ☐ cm = $\frac{9}{10}$ m

e. 4.25 m = ☐ cm = ☐ m

f. ☐ m = 312 cm = ☐ m

3. Draw and color an array to match each equation. Then write the product.

a. $\frac{2}{6} \times \frac{3}{4} = \frac{\ }{\ }$

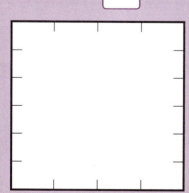

b. $\frac{6}{8} \times \frac{1}{3} = \frac{\ }{\ }$

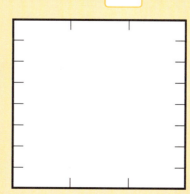

c. $\frac{4}{5} \times \frac{1}{4} = \frac{\ }{\ }$

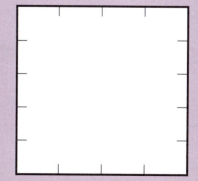

1. Loop the part that you would do first in each of these expressions.

a. 32 + 4 − 2
b. 32 × 4 + 2
c. 32 ÷ 4 + 2
d. 32 + 4 × 2
e. 16 ÷ 8 × 2
f. 16 ÷ 8 + 2
g. 16 − 8 × 2
h. 16 + 8 ÷ 2

2. Convert these lengths.

a. _____ cm = 380 mm
b. 265 cm = _____ mm
c. _____ cm = 90 mm
d. 72 cm = _____ mm
e. _____ cm = 85 mm
f. 46.2 cm = _____ mm
g. _____ cm = 128 mm
h. 138.1 cm = _____ mm

3. Complete each number sentence. Use the grid to help you.

a. $\frac{1}{5} \times \frac{1}{6} = $ ____

b. $\frac{7}{5} \times \frac{4}{6} = $ ____

c. $\frac{3}{5} \times \frac{8}{6} = $ ____

d. $\frac{8}{5} \times \frac{7}{6} = $ ____

e. $\frac{6}{5} \times \frac{11}{6} = $ ____

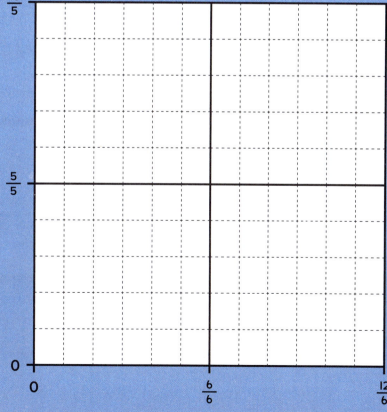

BIRTHPLACES

★ Figure out each of these and write the answer. Then write each letter above its matching answer at the bottom of the page. Some letters appear more than once.

16.2 + 9 = ___ **i**	19 − 16.7 = ___ **a**	3.7 + 24.2 = ___ **o**
23.1 − 16.2 = ___ **n**	0.7 + 29.6 = ___ **j**	15 − 7.2 = ___ **r**
25.8 + 0.9 = ___ **m**	34.5 − 7 = ___ **h**	17.8 + 5.4 = ___ **f**
12.4 − 0.7 = ___ **y**	4.9 + 10.5 = ___ **w**	18.8 − 8.3 = ___ **t**
8.3 + 18.8 = ___ **s**	21.7 − 2.8 = ___ **e**	12 + 9.8 = ___ **p**
30.2 − 1.5 = ___ **d**	20.4 + 10.7 = ___ **b**	30.4 − 0.8 = ___ **l**
59.1 + 1.2 = ___ **c**		

30.3 25.2 26.7 26.7 11.7   60.3 2.3 7.8 10.5 18.9 7.8

15.4 2.3 27.1   10.5 27.5 18.9   23.2 25.2 7.8 27.1 10.5

21.8 7.8 18.9 27.1 25.2 28.7 18.9 6.9 10.5   31.1 27.9 7.8 6.9

25.2 6.9 2.3

27.5 27.9 27.1 21.8 25.2 10.5 2.3 29.6

1. Calculate each perimeter. Record the steps you use.

a.

_____ m

b.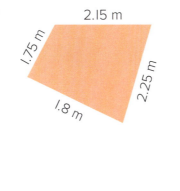

_____ m

c.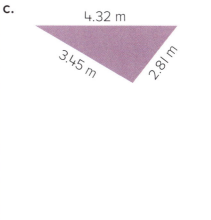

_____ m

2. Use a metric ruler to measure each strip.
   Write the length in millimeters and as a decimal fraction of a meter.

a.

_____ mm   _____ m

b.

_____ mm   _____ m

c.

_____ mm   _____ m

3. Complete each number sentence. Use the grid to help you.

a. $\frac{2}{3} \times 1\frac{1}{4} =$ ☐

b. $1\frac{1}{3} \times 2\frac{3}{4} =$ ☐

c. $1\frac{2}{3} \times 1\frac{1}{2} =$ ☐

d. $1\frac{1}{3} \times 2\frac{1}{4} =$ ☐

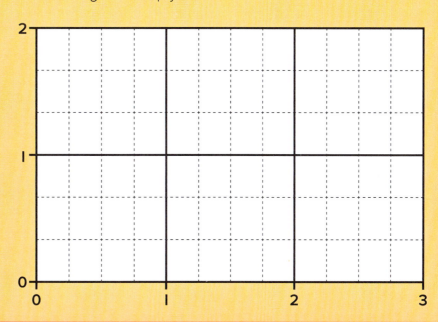

1. Complete each equation. Show your thinking.

a. $4\frac{4}{6} - 1\frac{5}{6} =$ ☐

b. $\frac{6}{9} - \frac{1}{3} =$ ☐

c. $\frac{2}{3} - \frac{3}{5} =$ ☐

2. Write each length as a decimal fraction then as a mixed number.

a. 4,355 m
_____ km
$4\frac{355}{1000}$ km

b. 2,350 m
2.35 km
_____ km

c. 3,465 m
_____ km
_____ km

d. 5,295 m
_____ km
_____ km

3. For each of these, complete the diagram to compare the amounts. Write an equation to match then complete the sentence.

a. A $\frac{1}{2}$
   B [bar divided into parts] 2 yd
   ___ × ___ = ___

b. A $\frac{1}{4}$
   B $2\frac{1}{2}$ ft
   ___ × ___ = ___

c. A 4
   B 24
   ___ × ___ = ___

d. A $\frac{1}{3}$
   B 3
   ___ × ___ = ___

FLYING HIGH

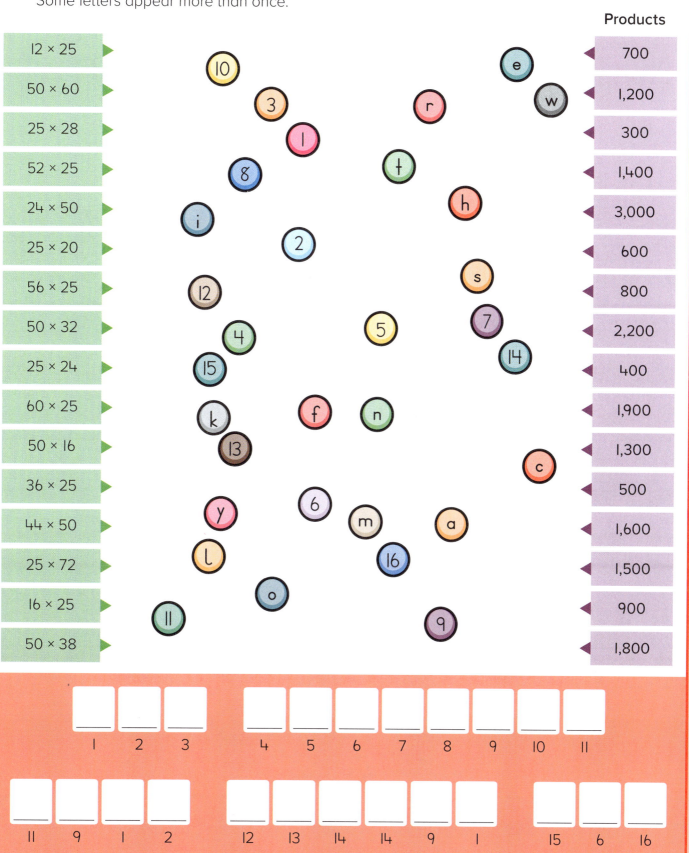

1. Use a written method to calculate the difference between the amount in the wallet and the price.

a. $17.35 $12.95

$ _____

b. $7.90 $15.45

$ _____

c. $6.79 $10.65

$ _____

2. Solve each problem. Write your answer two ways. Show your thinking.

a. The perimeter of a rectangle is 68 cm. The longer sides are 230 mm each. How long is each other side?

_____ mm   _____ cm

b. The swimming pool is 25 m long. Daniel swims 40 lengths each week. How far does he swim each week?

_____ km   _____ m

3. Solve the problem. Show your thinking.

James wants to lay some rubber tiles in his home gym. The room measures $12\frac{1}{2}$ ft by $15\frac{1}{3}$ ft. The tiles cost $6 a square foot. What is the area of the room and the cost of the rubber tiles?

Area _____ sq ft   Total cost $ _____

1. Write the matching number in words.

   a.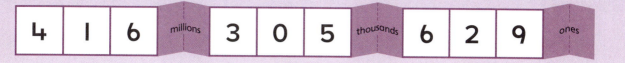

   _____

   _____

   _____

   b.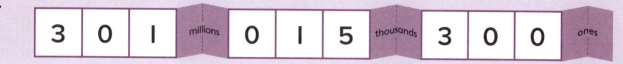

   _____

   _____

   _____

2. Calculate each product. Show your thinking.

   a. $3 \times 2\frac{3}{4}$

   b. $2 \times 4\frac{3}{5}$

   c. $6 \times 1\frac{1}{4}$

   d. $3 \times 1\frac{4}{6}$

3. Write the products. Color or outline parts of the squares to show your thinking.

   a. $4 \times 0.2 = $ _____
   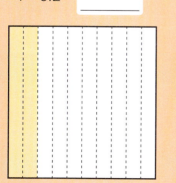

   b. $2 \times 0.3 = $ _____
   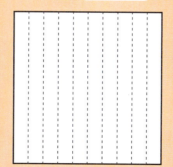

   c. $6 \times 0.1 = $ _____
   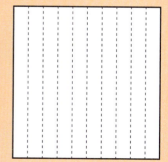

MATCHING?

What do storms, needles, and potatoes have in common?

★ Figure out each of these and write the product. Then write each letter above its matching product at the bottom of the page.

12 × 38 = ___ y	24 × 12 = ___ l	18 × 11 = ___ u
31 × 13 = ___ n	12 × 42 = ___ e	13 × 41 = ___ a
21 × 11 = ___ h	26 × 12 = ___ e	33 × 11 = ___ s
11 × 42 = ___ t	14 × 18 = ___ s	12 × 32 = ___ h
13 × 15 = ___ t	45 × 11 = ___ o	22 × 14 = ___ a
43 × 12 = ___ b	26 × 11 = ___ a	13 × 23 = ___ e
14 × 25 = ___ e	12 × 18 = ___ v	11 × 41 = ___ e
34 × 12 = ___ n	13 × 35 = ___ l	45 × 14 = ___ t
11 × 17 = ___ y	12 × 25 = ___ c	14 × 34 = ___ e

___ ___ ___ ___   ___ ___ ___   ___ ___ ___ ___
195 231 350 456   286 455 288   384 533 216 312

___ ___ ___ ___   ___ ___ ___
451 187 504 252   516 198 462

___ ___ ___ ___ ___ ___   ___ ___ ___
300 308 408 403 495 630   363 299 476

1. **a.** Write these decimal fractions in order from **least** to **greatest**. Use the number line to help you.

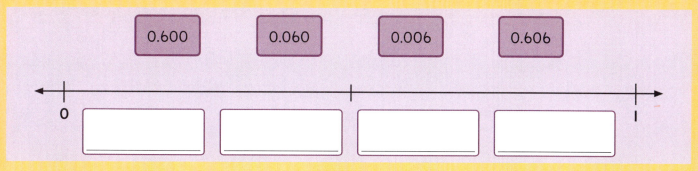

b. Write these in order from **greatest** to **least**.

| 0.315 | 0.153 | 0.531 | 1.053 | 1.305 | 0.135 |

_____ _____ _____ _____ _____ _____

2. Draw and color an array to match each equation. Then write the product.

a. $\frac{2}{5} \times \frac{2}{3} = \underline{}$

b. $\frac{3}{4} \times \frac{1}{3} = \underline{}$

c. $\frac{4}{6} \times \frac{1}{4} = \underline{}$

3. Complete each equation. Draw jumps on the number line to show your thinking.

a. $6 \times 0.2 = \underline{}$

b. $5 \times 0.3 = \underline{}$

c. $3 \times 0.2 = \underline{}$

1. Complete these calculations. Show your thinking.

   a. $736 \div 4 = \underline{}$

   b. $2{,}343 \div 3 = \underline{}$

   c. $1{,}377 \div 9 = \underline{}$

2. Convert ounces to pounds to complete these.

   a. 18 oz = ☐ lb ☐ oz
   b. 20 oz = ☐ lb ☐ oz
   c. 28 oz = ☐ lb ☐ oz
   d. 34 oz = ☐ lb ☐ oz
   e. 37 oz = ☐ lb ☐ oz
   f. 40 oz = ☐ lb ☐ oz
   g. 69 oz = ☐ lb ☐ oz
   h. 57 oz = ☐ lb ☐ oz
   i. 73 oz = ☐ lb ☐ oz

3. Multiply the ones and then the tenths to figure out each partial product. Then write the total.

   a. buy 3 boxes — 4.2 kg
   $(\underline{3} \times \underline{4}) + (\underline{3} \times \underline{0.2})$
   $\underline{} + \underline{} = \underline{}$ kg

   b. buy 5 boxes — 3.4 kg
   $(\underline{} \times \underline{}) + (\underline{} \times \underline{})$
   $\underline{} + \underline{} = \underline{}$ kg

   c. buy 6 boxes — 5.3 kg
   $(\underline{} \times \underline{}) + (\underline{} \times \underline{})$
   $\underline{} + \underline{} = \underline{}$ kg

   d. buy 5 boxes — 6.5 kg
   $(\underline{} \times \underline{}) + (\underline{} \times \underline{})$
   $\underline{} + \underline{} = \underline{}$ kg

VACATION FUN

Jamal went camping and got something. He sat down to search for it and came home without finding it. What is it?

★ Figure out each of these and write the quotient. Then find each quotient in the grid below and cross out the letter above. Write the remaining letters at the bottom of the page.

738 ÷ 6 = 123	296 ÷ 8 = 37	864 ÷ 9 = 96
472 ÷ 8 = 59	378 ÷ 9 = 42	996 ÷ 6 = 166
576 ÷ 9 = 64	888 ÷ 6 = 148	656 ÷ 8 = 82
456 ÷ 8 = 57	992 ÷ 8 = 124	288 ÷ 9 = 32
744 ÷ 8 = 93	198 ÷ 9 = 22	378 ÷ 6 = 63
675 ÷ 9 = 75	564 ÷ 6 = 94	576 ÷ 8 = 72
852 ÷ 6 = 142	368 ÷ 8 = 46	477 ÷ 9 = 53
184 ÷ 8 = 23	774 ÷ 9 = 86	282 ÷ 6 = 47

H	A	T	S	C	R	E	W	C	A	P
148	34	22	58	123	32	64	75	59	142	44
B	L	I	S	T	E	R	N	E	S	T
42	92	136	94	82	47	53	160	23	57	95
S	H	E	L	T	E	R	B	U	R	N
124	46	37	72	86	62	63	93	96	24	166

Write the letters in order from the ✻ to the bottom-right corner.

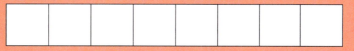

A SPLINTER

1. Use the standard division algorithm to calculate each quotient. Remember to estimate before or after your calculation to check your accuracy.

a. T O	b. T O	c. T O	d. T O
4) 6 4	3) 5 1	5) 7 5	2) 6 8

e. T O	f. T O	g. T O	h. T O
6) 9 6	3) 9 9	6) 7 8	3) 5 7

2. Write the missing mass to make each balance picture true.

a. 3 lb | 34 oz ____ oz

b. 18 oz ____ oz | 1.75 lb

c. 21 oz ____ oz | 2 lb

d. 1.5 lb | 14 oz ____ oz

3. Write the products. Color or outline parts of the squares to show your thinking.

a. 5 × 0.05 = ____

b. 4 × 0.12 = ____

c. 3 × 0.25 = ____

1. Complete these calculations using the standard division algorithm.

a.
Th	H	T	O

4) 6 8 9 6

b.
Th	H	T	O

4) 2 4 9 6

c.
Th	H	T	O

5) 3 7 5 5

2. Solve these word problems. Show your thinking.

a. Mom split 3 lb of ground meat into 2 separate bags. How much meat is in each bag?

_____ oz

b. Dad bought $2\frac{1}{2}$ lb of flour and 12 oz of sugar. What was the total weight of his purchase?

_____ oz

3. Figure out the total cost of each purchase.

a. buy 3 boxes $4.30

(_3_ × _4_) + (_3_ × _0.30_)

_____ + _____ = $ _____

b. buy 4 boxes $2.40

(___ × ___) + (___ × ___)

_____ + _____ = $ _____

c. buy 5 boxes $6.25

(___ × ___) + (___ × ___)

_____ + _____ = $ _____

d. buy 6 boxes $3.15

(___ × ___) + (___ × ___)

_____ + _____ = $ _____

GOLD MEDALS

What does every winner lose in a triathlon?

★ Figure out each of these and write the answer. Then find each answer in the grid below and cross out the letter above. Write the remaining letters at the bottom of the page. Write the letters in order from the ❋ to the bottom-right corner.

1.62 + 9.3 = _____

2.31 − 1.64 = _____

2.58 + 0.9 = _____

2.96 − 0.7 = _____

5.91 + 1.2 = _____

1.62 − 0.09 = _____

4.73 + 1.91 = _____

2.04 − 1.07 = _____

2 − 0.95 = _____

3.75 + 2.35 = _____

8 − 1.67 = _____

1.7 + 2.96 = _____

1.24 − 0.38 = _____

1.53 + 7.2 = _____

3.02 − 1.5 = _____

3.08 + 1.07 = _____

4.73 − 1.91 = _____

2.8 + 2.17 = _____

1.24 + 0.77 = _____

3.04 − 0.8 = _____

3.7 + 2.42 = _____

3.45 − 0.7 = _____

6.9 + 1.37 = _____

7.7 − 2.99 = _____

0.82 + 1.88 = _____

1.09 − 0.49 = _____

2.04 + 1.07 = _____

4.9 − 1.05 = _____

3.22 + 3.92 = _____

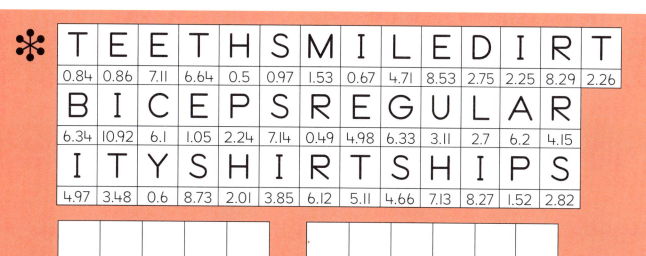

1. Write each length as a fraction of a meter. Use decimal fractions and common fractions.

a. 500 mm = 0.__5__ m = ☐ m

b. 10 mm = 0.____ m = $\frac{1}{100}$ m

c. 5 mm = 0.____ m = ☐ m

d. 50 mm = 0.____ m = ☐ m

e. 15 mm = 0.____ m = ☐ m

f. 250 mm = 0.____ m = ☐ m

2. This line plot show the mass of 20 tiger cubs born over a 12-month period.

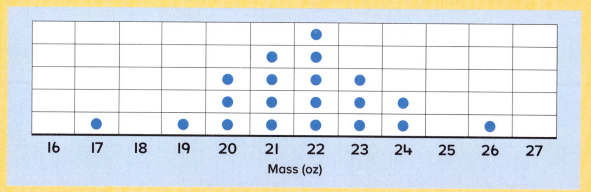

Mass (oz)

What does the shape of the line plot tell you about the mass of the cubs?

_____
_____

3. Write the products. Shade parts of the squares to match.

a. 0.5 × 0.3 = ____

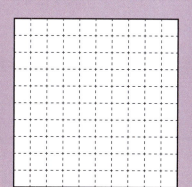

b. 0.8 × 0.4 = ____

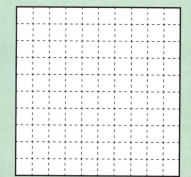

c. 0.6 × 0.8 = ____

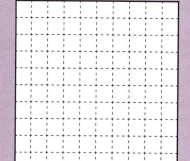

1. Calculate the total mass of these packages.

a. 27.5 kg, 3.65 kg

T O t h

+

b. 14.09 kg, 3.7 kg

T O t h

+

c. 21.85 kg, 3.4 kg

T O t h

+

2. Complete these equations. Use a pattern to help you.

a.
$4 \times 6 =$ ___
$4 \times 0.6 =$ ___
$4 \times 0.06 =$ ___

b.
$3 \times 7 =$ ___
$3 \times 0.7 =$ ___
$3 \times 0.07 =$ ___

c.
$8 \times 9 =$ ___
$8 \times 0.9 =$ ___
$8 \times 0.09 =$ ___

d.
$7 \times 8 =$ ___
$7 \times 0.8 =$ ___
$7 \times 0.08 =$ ___

e.
$9 \times 3 =$ ___
$9 \times 0.3 =$ ___
$9 \times 0.03 =$ ___

f.
$6 \times 2 =$ ___
$6 \times 0.2 =$ ___
$6 \times 0.02 =$ ___

3. Write the missing numbers to show how you would solve the problem. Color the diagram to show each share.

Three oranges are shared equally by 8 players. How much of one whole orange will be in each share?

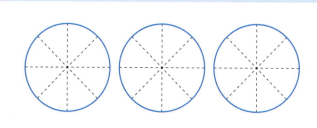

___ oranges shared by

___ players

$3 \div$ ___ $=$ ___

STAND OUT

Why do firefighters wear red suspenders?

★ Figure out each of these and write the product. Then find each product in the puzzle below and color its matching letter.

3.7 × 4 = ____
5.7 × 3 = ____
2 × 4.8 = ____

5 × 3.8 = ____
6 × 5.8 = ____
5.6 × 4 = ____

6.4 × 3 = ____
2 × 3.6 = ____
5 × 4.7 = ____

7 × 4.3 = ____
6.3 × 4 = ____
7.8 × 3 = ____

5.3 × 3 = ____
8 × 3.8 = ____
4 × 5.4 = ____

3 × 8.9 = ____
6.7 × 2 = ____
2.7 × 9 = ____

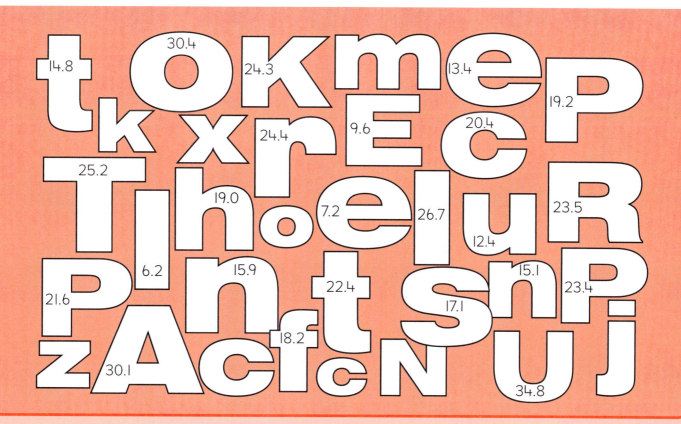

1. Complete these statements.

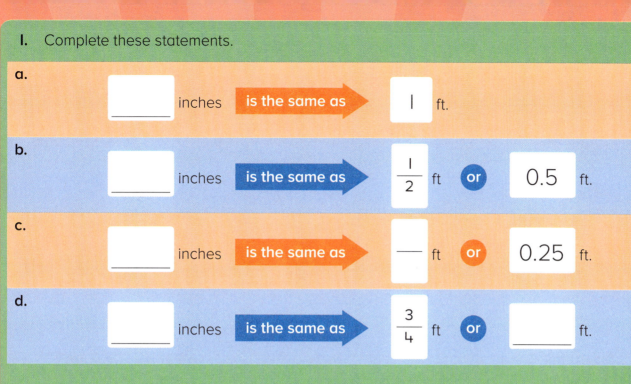

a. ____ inches is the same as 1 ft.

b. ____ inches is the same as $\frac{1}{2}$ ft or 0.5 ft.

c. ____ inches is the same as ___ ft or 0.25 ft.

d. ____ inches is the same as $\frac{3}{4}$ ft or ____ ft.

2. Figure out each partial product. Then write the total.

a. 3 × 4.15

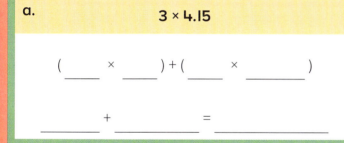

(___ × ___) + (___ × ___)

___ + ___ = ___

b. 4 × 6.12

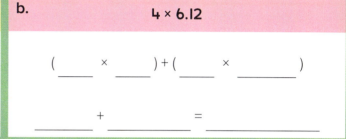

(___ × ___) + (___ × ___)

___ + ___ = ___

c. 5 × 7.05

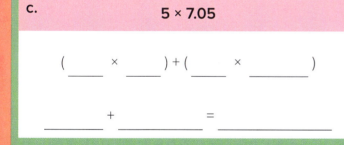

(___ × ___) + (___ × ___)

___ + ___ = ___

d. 6 × 5.35

(___ × ___) + (___ × ___)

___ + ___ = ___

3. Each circle below is one whole. Show the fraction in two different ways. Then complete the sentence.

$\frac{3}{4}$ is the same as ___ divided by ___ .

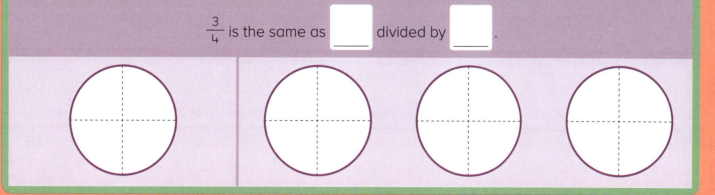

1. Figure out the difference. Draw jumps to show your thinking.

   $8.2 - 5.7 =$ _____

2. Write the products. Shade parts of the squares to match.

a. $0.7 \times 0.6 =$ _____
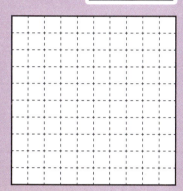

b. $0.3 \times 0.9 =$ _____
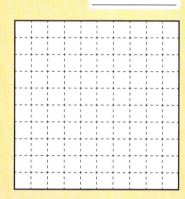

c. $0.8 \times 0.4 =$ _____
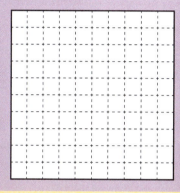

d. $0.7 \times 0.3 =$ _____
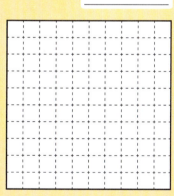

e. $0.5 \times 0.9 =$ _____
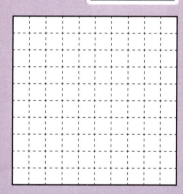

f. $0.6 \times 0.4 =$ _____
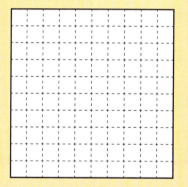

3. Each square is one whole. The striped part shows the fraction. Draw more lines and color parts to divide the striped part by the whole number. Then complete the equation.

a.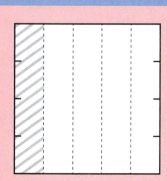

$\dfrac{1}{5} \div 4 =$ _____

b.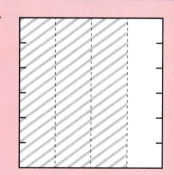

$\dfrac{3}{4} \div 6 =$ _____

RUG UP

What kind of coat can you put on only when it is wet?

★ Figure out each of these and draw a straight line to the correct product. The line will pass through a number and a letter. Write the letter above its matching number at the bottom of the page.

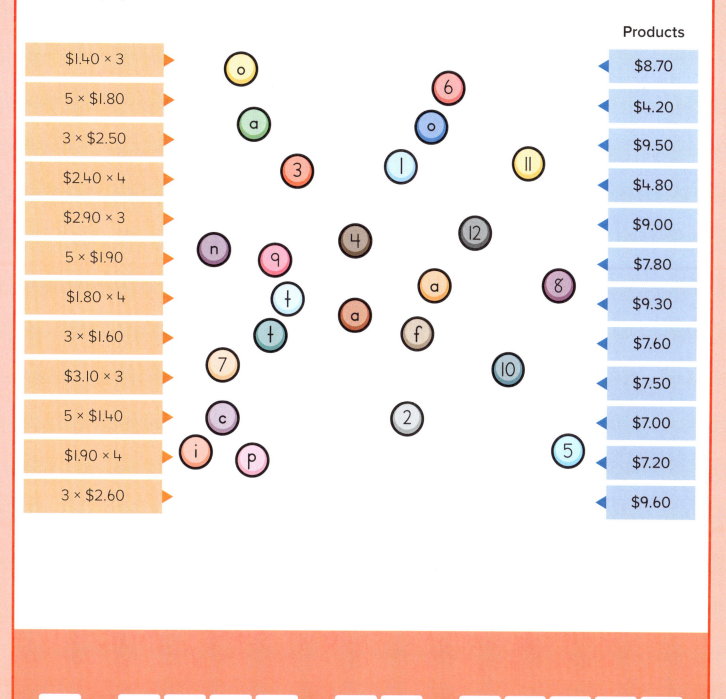

1. Figure out each difference. Show your thinking.

a.
36.4 − 12.52 = _____

b.
15.06 − 9.5 = _____

c.
19.47 − 8.09 = _____

2. Write the common fractions that you could multiply to figure out each product. Then complete the equations.

a.
0.6 × 0.03 = _____

___ × ___ = ___

b.
0.7 × 0.08 = _____

___ × ___ = ___

c.
0.04 × 0.7 = _____

___ × ___ = ___

d.
0.09 × 0.3 = _____

___ × ___ = ___

3. Color the ○ to show whether the problem is multiplication or division. Then figure out the solution to each problem. Show your thinking.

a. Two people equally share $\frac{1}{4}$ gallon of water. How much water will be in each share?

○ Multiplication   ○ Division

_____ gallon

b. Four hamsters weigh $\frac{1}{3}$ of a pound in total. If they are all about the same size, how much does each hamster weigh?

○ Multiplication   ○ Division

_____ pound

1. Write numbers to show equal lengths. Use decimal fractions and mixed numbers where necessary.

a. 0.5 m = ____ cm = $\frac{1}{2}$ m

b. ____ mm = 7.5 cm = ____ cm

c. ____ mm = ____ cm = $12\frac{1}{2}$ cm

d. 1.25 m = ____ cm = ____ cm

e. 3 m = ____ cm = ____ m

f. ____ mm = ____ cm = $\frac{2}{10}$ cm

2. Draw a red needle to show the mass on the scale. Then write the mass two other ways.

a. 0.8 kg

b. 1.5 kg

c. 3,200 g

d. $2\frac{1}{5}$ kg

e. 4,500 g

f. 2.75 kg

3. Complete these. You can draw lines in the diagram to help.

Shade 5 whole shapes. How many fourths are in 5?

____ × $\frac{1}{4}$ = 5

5 ÷ $\frac{1}{4}$ = ____

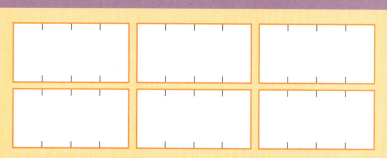

MATH QUIZ

★ These students were given a math quiz. Check their answers and draw a ✔ beside each correct answer. Add the ✔ for each student and write the score at the bottom of each paper.

Name: Corey

1. $(280 - 40) \div 6 =$ 40
2. $60 \times 15 \div 3 =$ 30
3. $32 \div 8 + 8 =$ 2
4. $120 - 4 \times 21 =$ 36
5. $3 \times 20 \div 5 =$ 12
6. $7 + 150 \div 3 =$ 52
7. $240 \div (23 + 17) =$ 6
8. $64 - 32 \div 8 =$ 4
9. $(25 + 11) \div 6 =$ 6
10. $7 \times 7 - 7 =$ 42
11. $8 \times 11 - 5 =$ 83
12. $8 \times (11 - 5) =$ 48
13. $180 \div 90 \times 9 =$ 18
14. $48 \div 12 - 2 =$ 2
15. $(50 - 2) \div 12 =$ 4
16. $96 \div 3 \times 2 =$ 64
17. $77 \div (7 + 4) =$ 7
18. $77 \div 7 + 4 =$ 7
19. $11 \times (11 - 9) =$ 22
20. $3 \times 8 + 4 =$ 36

Total correct: _____

Name: Eva

1. $(280 - 40) \div 6 =$ 40
2. $60 \times 15 \div 3 =$ 30
3. $32 \div 8 + 8 =$ 12
4. $120 - 4 \times 21 =$ 1392
5. $3 \times 20 \div 5 =$ 12
6. $7 + 150 \div 3 =$ 57
7. $240 \div (23 + 17) =$ 6
8. $64 - 32 \div 8 =$ 60
9. $(25 + 11) \div 6 =$ 6
10. $7 \times 7 - 7 =$ 42
11. $8 \times 11 - 5 =$ 83
12. $8 \times (11 - 5) =$ 48
13. $180 \div 90 \times 9 =$ 18
14. $48 \div 12 - 2 =$ 3
15. $(50 - 2) \div 12 =$ 4
16. $96 \div 3 \times 2 =$ 16
17. $77 \div (7 + 4) =$ 7
18. $77 \div 7 + 4 =$ 15
19. $11 \times (11 - 9) =$ 22
20. $3 \times 8 + 4 =$ 28

Total correct: _____

Who won the quiz? _____

1. Complete each number sentence. Use the grid to help you.

a. $1\frac{2}{5} \times 2\frac{3}{4} =$ ☐

b. $1\frac{3}{5} \times \frac{1}{2} =$ ☐

c. $1\frac{4}{5} \times 1\frac{1}{4} =$ ☐

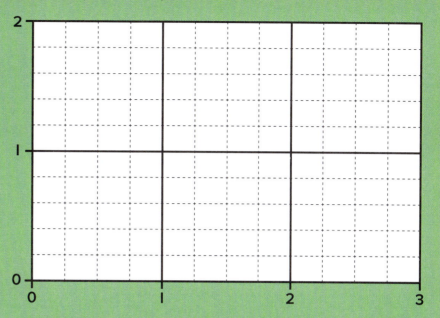

2. These are weights of different watermelons. Record each weight in the line plot below. You will need to convert the grams to kilograms.

7,500 g	$5\frac{1}{2}$ kg	3 kg	6 kg	$2\frac{1}{2}$ kg	5,500 g
5 kg	2,500 g	4,500 g	2 kg	$5\frac{1}{2}$ kg	6 kg
$4\frac{1}{2}$ kg	5 kg	5,500 g	$6\frac{1}{2}$ kg	5 kg	$3\frac{1}{2}$ kg

Weight of Watermelons

Kilograms (kg): 2, $2\frac{1}{2}$, 3, $3\frac{1}{2}$, 4, $4\frac{1}{2}$, 5, $5\frac{1}{2}$, 6, $6\frac{1}{2}$, 7, $7\frac{1}{2}$, 8

3. Convert each amount.

a. 2 gal = ☐ qt

b. 4.5 gal = ☐ qt

c. 6.75 gal = ☐ qt

d. ☐ gal = 4 qt

e. ☐ gal = 16 qt

f. ☐ gal = 22 qt

1. Complete these statements.

a. _____ ounces is the same as 1 lb.

b. _____ ounces is the same as $\frac{1}{2}$ lb or 0.5 lb.

c. _____ ounces is the same as $\frac{1}{4}$ lb or _____ lb.

d. _____ ounces is the same as ___ lb or 0.75 lb.

2. Complete these.

a. Shade 4 whole shapes. How many thirds are in 4?

_____ × $\frac{1}{3}$ = 4

4 ÷ $\frac{1}{3}$ = _____

b. Shade 3 whole shapes. How many fifths are in 3?

_____ × $\frac{1}{5}$ = 3

3 ÷ $\frac{1}{5}$ = _____

3. Solve each problem. Show your thinking.

a. A set of 6 dining chairs costs $441. What is the cost of each chair?

$ _____

b. Four friends share the cost of buying a dining suite for $441, and a lounge suite for $526. What is each person's share?

$ _____

ORIGO Stepping Stones 5 · 12.2

ANNOYING INSECTS

What flies around all day but never goes anywhere?

 Figure out each of these and write the product. Find the product in the grid below and cross out the letter above. Then write the remaining letters at the bottom of the page.

$1.75 × 3 = $ _____

4 × $1.95 = $ _____

3 × $2.65 = $ _____

4 × $1.65 = $ _____

$1.15 × 5 = $ _____

$1.35 × 3 = $ _____

$3.15 × 3 = $ _____

4 × $2.45 = $ _____

5 × $1.85 = $ _____

4 × $2.25 = $ _____

$1.55 × 3 = $ _____

$1.25 × 5 = $ _____

$1.55 × 5 = $ _____

5 × $1.45 = $ _____

4 × $2.15 = $ _____

5 × $1.95 = $ _____

$2.25 × 3 = $ _____

$1.75 × 4 = $ _____

$2.45 × 3 = $ _____

Write the letters in order from the ✻ to the bottom-right corner.

A	S	A	N	D	F	L	Y
$5.25	$6.25	$4.55	$9.45	$7.75	$7.05	$9.75	$6.60
F	L	I	E	S	B	E	E
$7.35	$4.15	$7.80	$9.80	$6.75	$7.95	$7.25	$9.25
G	N	A	T	P	I	G	S
$5.75	$7.00	$5.95	$4.65	$8.60	$4.05	$9.55	$9.00

 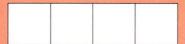

ORIGO Stepping Stones 5 • 12.3

1. Complete these equations. Draw jumps on the number line to show your thinking.

a. 6 × 0.08 = _____

b. 4 × 0.09 = _____

c. 5 × 0.07 = _____

2. Solve each problem. Show your thinking.

a. A recipe requires $\frac{1}{4}$ of a cup of almonds. How many batches can be made with 3 cups of almonds?

_____ batches

b. A recipe requires $2\frac{1}{3}$ cups of plain flour and $\frac{1}{3}$ cup of sugar. How much flour is required for 3 batches?

_____ cups of flour

3. Break each number into parts that you can easily divide. Calculate the partial quotients. Then complete the equations.

a.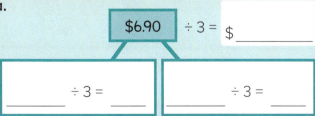

b.

c.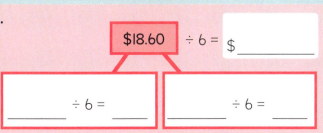

d.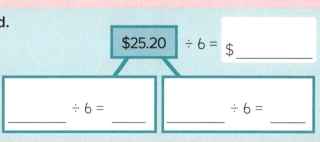

ORIGO Stepping Stones 5 • 12.4

102

1. Figure out each partial product. Then write the total.

a.
4 × 0.18 = _____
4 × 0.1 = _____
4 × 0.08 = _____

b.
6 × 0.12 = _____
6 × 0.1 = _____
6 × 0.02 = _____

c.
3 × 0.27 = _____
3 × 0.2 = _____
3 × 0.07 = _____

2. Write the missing amount so that each balance picture is true.

a.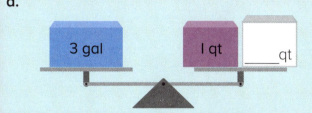

b.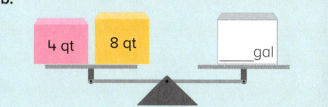

c.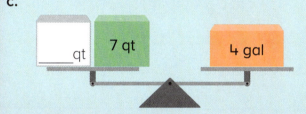

d.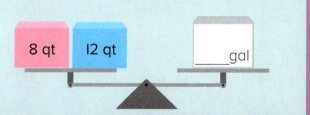

3. Read the equation. Then complete the picture to help you write the answer.

a. 4 ÷ 0.2 = _____

× 4   ◯ × 0.2 = 1   × 4
      ◯ × 0.2 = 4

b. 6 ÷ 0.3 = _____

× 2   ◯ × 0.3 = 3   × 2
      ◯ × 0.3 = 6

c. 8 ÷ 0.5 = _____

d. 4 ÷ 0.4 = _____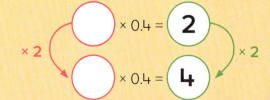

DATE LINE

Which month has 28 days each year and 29 days every leap year?

★ Figure out each of these and write the answer. Then find each answer in the puzzle below and shade the matching letter. The remaining letters will spell the answer.

21 − 9.6 =	37.7 + 0.8 =	20.3 − 6.8 =
9.7 + 21.6 =	25.1 − 9.6 =	25.1 + 9.6 =
17.4 − 8.7 =	9.6 + 6.9 =	14.8 − 7.9 =
7.9 + 14.8 =	35 − 10.8 =	0.9 + 20.9 =
26.1 − 0.8 =	3.3 + 7.7 =	42 − 8.6 =
13.5 + 18.6 =	18.6 − 13.7 =	6.8 + 20.3 =
9.6 − 6.9 =	8.7 + 17.4 =	25.1 − 7.6 =

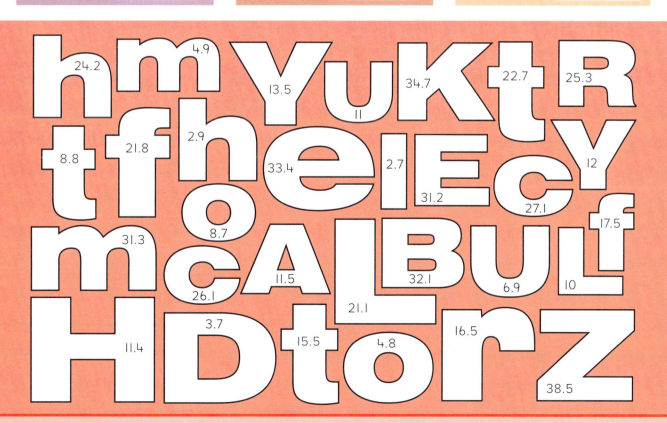

ORIGO Stepping Stones 5 • 12.7   104

1. Write the products. You can shade parts of each square to help your thinking.

a. $3 \times 0.15 =$ _____
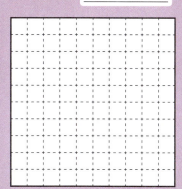

b. $4 \times 0.14 =$ _____
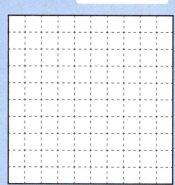

c. $3 \times 0.28 =$ _____
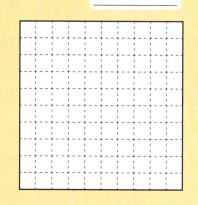

2. Convert each amount to fluid ounces. Write number sentences to show your thinking.

a. 6 qt = _____ fl oz

b. 3.25 qt = _____ fl oz

3. Write the answers in these place-value charts.

a. **Multiplication**

	H	T	O	t	h
$6 \times 30 =$				.	
$6 \times 3 =$				.	
$6 \times 0.3 =$				.	
$6 \times 0.03 =$				.	

b. **Division**

	H	T	O	t	h
$6 \div 30 =$				.	
$6 \div 3 =$				.	
$6 \div 0.3 =$				.	
$6 \div 0.03 =$				.	

1. Estimate the products then complete each equation. Show your thinking.

a. Estimate _____
4 × 3.35 = _____

b. Estimate _____
6 × 5.60 = _____

c. Estimate _____
5 × 6.75 = _____

2. Convert each amount to fluid ounces. Write number sentences to show your thinking.

a. 4 gal = _____ fl oz

b. 2.5 gal = _____ fl oz

3. Complete these statements.

a. 8 ÷ 0.4
is the same as
____ tenths ÷ ____ tenths
which is the same as
80 ÷ ____ = ____

b. 7 ÷ 0.2
is the same as
____ tenths ÷ ____ tenths
which is the same as
70 ÷ ____ = ____

c. 6 ÷ 0.5
is the same as
____ tenths ÷ ____ tenths
which is the same as
60 ÷ ____ = ____

d. 6 ÷ 0.3
is the same as
____ tenths ÷ ____ tenths
which is the same as
60 ÷ ____ = ____

e. 9 ÷ 0.3
is the same as
____ tenths ÷ ____ tenths
which is the same as
90 ÷ ____ = ____

f. 12 ÷ 0.6
is the same as
____ tenths ÷ ____ tenths
which is the same as
120 ÷ ____ = ____

NATIVE ANIMAL

What's big, white, furry, and found in the Grand Canyon?

★ Figure out each of these and write the answer. Then write each letter above its matching answer at the bottom of the page.

$3.48 ÷ 4 = $ ____ **r**	$8.40 ÷ 5 = $ ____ **o**
5 × $3.15 = $ ____ **l**	4 × $5.50 = $ ____ **a**
$6.24 ÷ 4 = $ ____ **a**	$7.20 ÷ 3 = $ ____ **y**
$6.50 × 4 = $ ____ **s**	$2.25 × 3 = $ ____ **e**
$7.50 ÷ 5 = $ ____ **a**	$8.70 ÷ 3 = $ ____ **r**
$5.25 × 4 = $ ____ **t**	5 × $4.49 = $ ____ **o**
$6.60 ÷ 3 = $ ____ **b**	$9.75 ÷ 5 = $ ____ **v**
5 × $3.20 = $ ____ **l**	$4.30 × 4 = $ ____ **p**
$4.80 ÷ 4 = $ ____ **r**	3 × $3.15 = $ ____ **e**

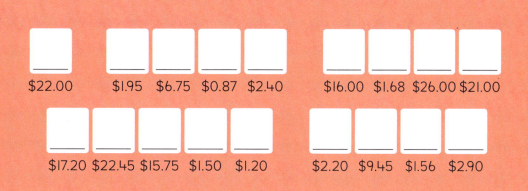

$22.00 $1.95 $6.75 $0.87 $2.40 $16.00 $1.68 $26.00 $21.00

$17.20 $22.45 $15.75 $1.50 $1.20 $2.20 $9.45 $1.56 $2.90

1. Complete each equation. Show the steps you use.

a. $3.2 \times 4.06 = \underline{\hspace{2cm}}$

b. $4.1 \times 2.8 = \underline{\hspace{2cm}}$

2. Solve these word problems. Show your thinking.

a. Mia's drink bottle holds 2 quarts. Jacob's bottle holds 40 fl oz. How much more does Mia's bottle hold than Jacob's bottle?

_____ fl oz

b. A leaking faucet loses 34 fl oz each day. How much water is lost after one week?

_____ gal _____ fl oz

3. Color the three labels that match the amount in each container.

a.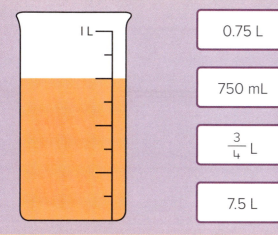

b.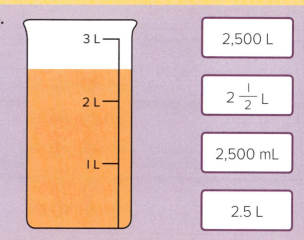

Working Space

Working Space